TK5103.45 GRO

DATE DUE FOR RETURN

CDMA Mobile Radio Design

For a listing of recent titles in the *Artech House Mobile Communications Series*, turn to the back of this book.

CDMA Mobile Radio Design

John B. Groe
Lawrence E. Larson

Artech House
Boston • London
www.artechhouse.com

Library of Congress Cataloging-in-Publication Data
Groe, John B.
   CDMA mobile radio design/ John B. Groe, Lawrence E. Larson.
      p. cm. — (Artech House mobile communications series)
   Includes bibliographical references and index.
   ISBN 1-58053-059-1 (alk. paper)
      1. Code division multiple access. 2. Cellular telephone systems. 3. Mobile
communication systems. I. Larson, Lawrence E. II. Title. III. Series.

TK5103.452.G76   2000   00-027455
621.3845—dc21   CIP

British Library Cataloguing in Publication Data
Groe, John B.
   CDMA mobile radio design. — (Artech House mobile
   communications series)
   1. Cellular radio—Design 2. Wireless communication systems
   —Design 3. Code division multiple access
   I. Title II. Larson, Lawrence E.
   621.3'845

   ISBN 1-58053-059-1

Cover design by Igor Valdman

© 2000 ARTECH HOUSE, INC.
685 Canton Street
Norwood, MA 02062

International Standard Book Number: 1-58053-059-1
Library of Congress Catalog Card Number: 00-027455

10 9 8 7 6 5 4 3 2

To my wife, Carmen; to our son, Matthew; and to my parents,
Jack and Dorothy.
—John B. Groe

To my wife, Carole; to our children, Hannah and Erik;
and to my mother, Jane.
—Lawrence E. Larson

Contents

Preface

Wireless communications is growing at a phenomenal rate. From 1991 to 1999, the number of subscribers increased from about 25 million to over 250 million. Incredibly, over the next seven years, the number of subscribers is expected to quadruple, to over 1 billion [1]. That growth rate is faster than that of any other consumer electronics product and is similar to that of the Internet.

Originally, wireless communications were motivated by and intended for mobile voice services. Later on, the first analog systems were improved with digital techniques, providing increased robustness and subscriber capacity. In the near future, digital systems will be augmented to try to meet users' insatiable need for even greater capacity and high-speed mobile data services.

Wireless communications rely on multiple-access techniques to share limited radio spectrum resources. These techniques, which use frequency, time, and power to divide the precious radio spectrum, are described in standards and are highly regulated. As such, infrastructure and subscriber manufacturers can be different and interchangeable.

This book details the complete operation of a mobile phone. It describes code division multiple access (CDMA) design issues but presents concepts and principles that are applicable to any standard. The book emphasizes CDMA because next-generation standards are based on that multiple-access technology.

This book uniquely ties together all the different concepts that form the mobile radio. Each of these concepts, in its own right, is suitable material for a book, if not several books, but is presented in such a way as to highlight key design issues and to emphasize the connection to other parts of the mobile radio.

Chapter 1 introduces some fundamentals of wireless communications. It describes the wireless network, which interfaces with landline services, and the procedures for communicating through the network. Chapter 1 illustrates the effects of radio propagation and reveals its impact on the mobile phone. It also lists some familiar wireless standards. Chapter 2 provides an overview of CDMA. It presents the basic concepts and highlights the key air interface requirements for the CDMA IS95 standard.

Chapter 3 introduces the digital system, which consists of a digital signal processor (DSP) and a microcontroller unit (MCU). The chapter uncovers the myriad of important roles the digital system plays. It also reviews some digital signal processing fundamentals and describes some tradeoffs in architecture. Chapter 4 introduces speech coding, a key function of the digital system. It shows how voice signals are translated to low bit rate data streams and vice versa. Chapter 5 provides detailed information about digital modulation and demodulaton. It presents a practical Rake receiver and describes the receiver's operation in the network. It also points out key timing issues and their effects on the performance of the mobile phone in the wireless network.

Chapter 6 describes data converters, circuits that interface the digital system to the auditory transducers (microphone and speaker) and the radio frequency (RF) transceiver. The chapter analyzes the nonideal effects of these interfaces and also presents fundamental data conversion techniques.

Chapter 7 is the first of three chapters dedicated to the RF transceiver, the mobile radio's connection to the air interface. It describes both the RF transmitter and the receiver from a system perspective, providing critical information about gain distribution and signal integrity. The chapter also presents insight into frequency synthesis and frequency planning in the mobile radio. Chapter 8 details the RF transmitter. It describes the transmit circuits between the digital-to-analog (D/A) converters' outputs and the antenna. The chapter covers the I/Q modulator, variable gain amplifier (VGA), up-converter, filters, driver, and power amplifier (PA). Chapter 9 details the operation of the RF receiver. It provides a circuit level view of the receiver from the antenna to the A/D converters' inputs. This chapter covers the low-noise amplifier (LNA), mixer, VGA, I/Q demodulator, and filters.

Chapter 10 describes next-generation wireless services and standards. The chapter points out improvements to CDMA IS95 that will accommodate more users and higher data rates. It also details leading next-generation CDMA proposals. Chapter 11 illustrates architecture advances to support improved CDMA IS95 performance and to meet the demands of next-generation CDMA networks. It addresses key areas, including the DSP, the RF transmitter, and the RF receiver.

A book covering such a range of systems, architectures, and circuits crosses several engineering disciplines. As a result, we benefited from discussions with and reviews by several colleagues. We would like to acknowledge Mr. Tom Kenney, Ryan Heidari, Sassan Ahmadi, and Ken Hsu of Nokia Mobile Phones; Professor George Cunningham of New Mexico Technical University; Professor Behzad Razavi of the University of California—Los Angeles; Professors Laurence Milstein, Peter Asbeck, Anthony Acompora, and Ian Galton of the University of California—San Diego; Professor John Long of the University of Toronto; and Mr. David Rowe of Sierra Monolithics.

Reference

[1] Viterbi, A. J., *CDMA: Principles of Spread-Spectrum Communications*, Reading, MA: Addison-Wesley, 1995.

1

Introduction to Wireless Communications

Wireless technology offers untethered service, newfound freedom, and the potential for "anytime, anyplace" communications. Consumers are embracing these services enthusiastically; their numbers are growing at a phenomenal rate and will continue to do so, as illustrated in Figure 1.1. The growth and the excitement of wireless communications are being driven by technological advancements that are making portable radio equipment smaller, cheaper, and more reliable. Those advancements include improved signal processing

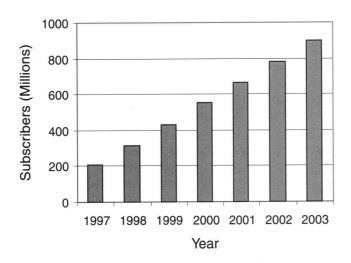

Figure 1.1 The growth rate of wireless subscribers is phenomenal [1].

1

techniques, innovative digital and radio frequency (RF) circuit design, and new large-scale integrated circuit methods.

This chapter introduces and describes key aspects of wireless networks. It investigates the wireline backbone, which facilitates wireless communications. That leads to an overview of the communication procedures used by both wireline and wireless networks. The chapter also details the effects of the radio link, which complicates radio design and leads to a variety of wireless standards.

1.1 Network Architecture for Cellular Wireless Communications

The wireless network supports over-the-air communications between mobile radios and stationary transceivers[1] known as base stations. These links are reliable only over short distances, typically tens of meters to a few kilometers. As such, a network of base stations is needed to cover a large geographic area, for example, a city. Base stations communicate through mobile switching centers, which connect to external networks such as the public telephone switching network (PTSN), the integrated services digital network (ISDN), and the Internet, as shown in Figure 1.2.

The mobile radio is free to move about the network. It relies on radio signals to form a wireless link to the base stations and therefore requires an RF transceiver. To support modern communication methods, the mobile radio

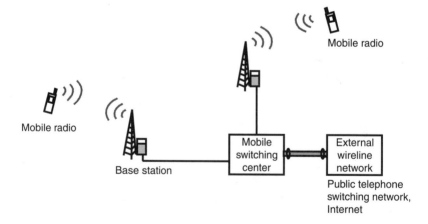

Figure 1.2 Wireless network architecture is an interconnection of mobile radios, base stations, mobile switching centers, and the external network.

---

1. Transmitter-receiver combinations.

includes a microcontroller unit (MCU) and a digital signal processor (DSP) to condition the signal before transmission and to demodulate the received signal (Figure 1.3).

The base stations translate the radio signals into data packets and signaling messages that are readable by the wireline network, which then forwards the information to the mobile switching center.

The mobile switching center routes the data packets based on the signaling messages and typically does not originate messages. In some cases, the mobile switching center may need to send queries to find wireless subscribers or portable local numbers (800- and 888-numbers).

The external network provides the communications backbone that connects the mobile switching centers. It routes data packets, screens messages for authorization, verifies routing integrity, and converts protocols. The external network may also act as a gateway to different networks.

The mobile switching center and the external network are signal transfer points that include measurement capabilities to indicate network problems and to monitor usage for billing purposes. Built-in redundancies in the network allow rerouting around faulty network points.

The network also includes service control points that interface to computers and provide database access. For example, the mobile switching center uses a service control point to access the home location register (HLR), the visitor location register (VLR), and the operation and maintenance center (OMC) files. Those databases list the subscribers in the home service area, track any

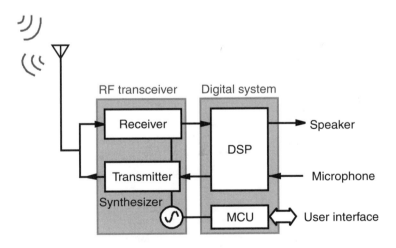

Figure 1.3 Modern mobile radio architecture consists of an RF transceiver and a digital system.

roaming (i.e., visiting) subscribers in the coverage area, and hold authentication files.

More information on network architectures can be found in [2–4].

1.2 Data Communication Techniques

Modern wireline and wireless networks rely on digital techniques for efficient communications. The techniques format message signals into data packets, thereby allowing multiple users to be "bundled" at higher network levels. That is important because it reduces the number of physical connections required to connect a set of users. The bundling occurs at signal transfer points and typically uses time multiplexing methods [2].

A basic wireline telephone channel for a single user supports a data rate of 64 Kbps; digital and optical data trunks carry higher data rates, as listed in Table 1.1.

The data packets are routed through the network by either circuit-switched or packet-switched connections. In circuit-switched networks, the path between the user and the destination node is set up at the time the connection is established, and any needed resources are reserved until the connection is terminated. In packet-switched networks, the path is not fixed but is dynamically selected based on network loading conditions and the destination address appended to each data packet.

Circuit-switched networks provide dedicated connections with low latency, while packet-switched networks offer greater flexibility with improved efficiency. Packet-switched networks are more complicated because data packets can take different paths and can be received out of order; the data packets must then be reassembled prior to final delivery to the user.

Table 1.1
Common Data Rates for Digital and Optical Networks [2]

Carrier Designation	Type	Bandwidth	Channels
DS0	Digital	64 Kbps	1
T-1	Digital	1.544 Mbps	24
T-3	Digital	44.736 Mbps	672
STM-1	Optical	51.84 Mbps	810
STM-3	Optical	155.52 Mbps	2,430
STM-16	Optical	2,488.32 Mbps	38,880

1.3 Protocols for Wireless Communications

Multiple users in communication networks are organized using routing and flow control procedures, known as protocols. A protocol is a set of rules governing data transmission and recovery in communication networks. The rules ensure reliable, seamless transmission of data and provide network management functions.

Protocols usually are organized as layers in a communication "stack." Data is passed up or down the stack one layer at a time, with specific functions performed at each layer.

Most communication networks follow the open system interconnections (OSI) model [5]. The seven-layer protocol stack, shown in Figure 1.4(a), includes the physical, data link, network, transport, session, presentation, and application layers. In wireless communication networks, a variation of the OSI model, the signaling system number 7 (SS7) model [2–3], is used. This four-level protocol stack, shown in Figure 1.4(b), mirrors the first three layers of the OSI model and combines the higher levels into a single application layer.

The protocol stack defines the architecture of each signal transfer point or node in the network. It uses the physical layer to interconnect those nodes and provide a path through the network, plus the data link and network layers to translate control signals and reformat data for communication with different

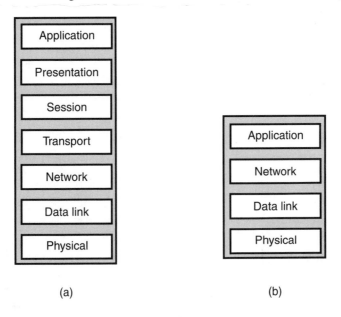

(a)　　　　　　　　　　　(b)

Figure 1.4 Network models: (a) OSI protocol stack typical of wireline networks and (b) SS7 protocol stack followed by wireless networks.

networks. Data always flows from one layer to the next in the protocol stack, as shown in Figure 1.5, to ensure robust communications.

Each layer in the protocol stack performs essential operations that are defined by the topology of the communication network. Those operations are outlined next.

The physical layer is the interface between two communication nodes. In a wireless network, the physical layer is the air interface between the mobile terminal and the base station. In a typical wireline network, it is the digital or optical trunk. The physical layer provides transfer services to higher layers in the protocol stack. Those transfer services use physical channels, also known as transport channels, with defined data rates, modulation schemes, power control methods, and RF parameters. The physical layer is different for each unique communication standard.

The data link layer combines the medium access control (MAC) and radio link control sublayers. The MAC sublayer maps basic functions known as logical channels to physical channels. That can be straightforward, or it can include multiplexing several logical channels onto a common physical channel. The data link layer also provides message sequencing, traffic monitoring, and signal routing to higher protocol layers.

The radio link control sublayer breaks down the data stream into data packets, also known as transport blocks, for transmission. It includes error control to ensure the integrity of the transmitted data. Typically, that means a parity check or a cyclic redundancy check (CRC) based on a polynomial generator [6]. The radio link control layer also interfaces with the higher protocol layers and provides call initialization, connection, and termination.

The network layer (or radio resource control layer) provides control and notification services. It supervises radio resources, including physical channel assignments, paging requests, and transmit power levels. It also interfaces to the wireline network and thereby enables connections to other users.

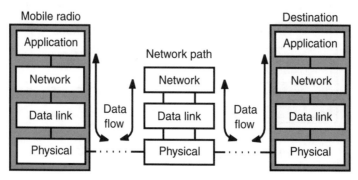

Figure 1.5 Data flow through the protocol stack for mobile communications.

The application layer represents the destination node. It specifies quality-of-service (QoS) requirements (priority levels, security, response time expectations, error rates, and recovery strategies) without the restrictions of the air and network interfaces. The application layer compresses and expands data in time to match the expectations of the mobile user.

The physical layer, the data link layer, and the network layer combine to form the message transfer part (MTP) of the SS7 protocol stack, as shown in Figure 1.6. The MTP of the SS7 model covers transmission from node to node in the communication network. It also interfaces with high-level protocols tailored to specific applications. For voice communications, one of two high-level protocols is used: the telephone user part (TUP) or the ISDN user part (ISUP).

1.4 Radio Propagation in a Mobile Wireless Environment

The radio interface is unique to wireless communications and is responsible for much of the complexity associated with wireless networks and mobile phones. The radio interface between the mobile phone and the base station is referred to as the communication channel and is affected by large- and small-scale factors. The large-scale effects are due to simple attenuation of the transmitted signal through the atmosphere. The small-scale effects behave unpredictably, vary sharply over small distances, and change quickly.

1.4.1 Path Loss

A transmitted signal is attenuated as it propagates through the atmosphere. This large-scale effect, known as path loss, is modeled by

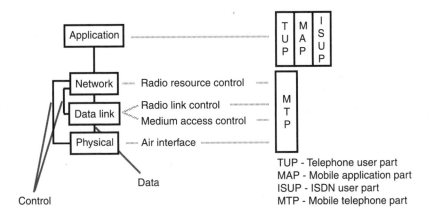

Figure 1.6 The SS7 model and the relationships among its constituent parts.

$$r(d) \propto d^{-n} \tag{1.1}$$

where $r(d)$ is the received power at a distance d separating the mobile and the base station, and n is the path loss exponent with typical values of 2.7 to 3.5 for urban cellular radio [7]. The model is quite simple and is appropriate only for line-of-sight propagation.

In practice, the signal path typically is cluttered by obstructions that reflect or block the transmitted signal and introduce statistical variability to the simple path loss model, as shown in Figure 1.7. This effect is known as shadowing and is modeled as a log-normal random variable [7]. That leads to a new expression for the received power:

$$r(d) \propto 10^{\chi/10} d^{-n} \tag{1.2}$$

where χ is the log-normal random variable used to model the shadowing effect.

1.4.2 Multipath Fading

The transmitted signal is not restricted to line-of-sight propagation. It can bounce off nearby obstructions, such as buildings and mountains, and arrive at the receiving antenna as shown in Figure 1.8. The reflected waves travel different paths to the receiving antenna and therefore experience different propagation delays and path losses. The resulting time-delayed versions of the signal are known as multipath rays. Multipath rays add vectorially and produce the fluctuations in the received power level shown in Figure 1.9, known as small-scale fading. Unfortunately, it is possible for multipath rays to combine

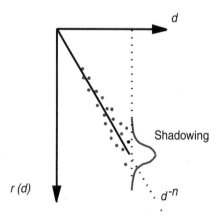

Figure 1.7 Received signal strength with path loss and log-normal shadowing.

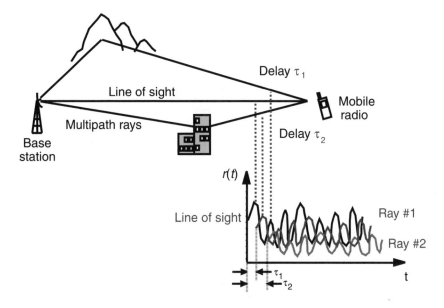

Figure 1.8 Multipath propagation of a transmitted signal arrives at the receiver with different delays.

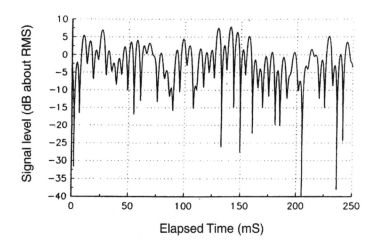

Figure 1.9 Multipath fading produces a wide variation in the received signal strength as a function of time in a mobile environment. (*From:* T. S. Rappaport, *Wireless Communications*, © 1995; reprinted by permission of Prentice-Hall, Inc., Upper Saddle River, NJ.)

destructively, and the received signal can disappear completely for a short period of time.

The effects of multipath fading combine with large-scale path losses to attenuate the transmitted signal as it passes through the channel, as shown in Figure 1.10. The graph shows that the received power level at a distance d from the transmitting antenna depends on the simple path loss model altered by the shadowing and multipath distributions.

Multipath fading is created by the frequency-selective and time-varying characteristics of the communication channel. Those characteristics are not deterministic and therefore must be analyzed using statistical methods. This approach is illustrated in the following examples.

In the first case, two sinusoidal signals at frequencies f_1 and f_2 are transmitted through the channel as shown in Figure 1.11. The signals are affected by the channel, which attenuates the power level, r, of each signal independently. The attenuation process for each signal varies with frequency and can be described by two distinct probability density functions (pdf's). If $f_1 \approx f_2$, then the pdf's of the received power levels $p(r)$ will be nearly the same, and the cross-correlation between the two, $R(\Delta f)$, will be high. As the separation between f_1 and f_2 increases, their amplitude pdf's will become dissimilar and their cross-correlation will be lower.

The coherence bandwidth, $(\Delta f)_c$, is the range of frequencies in which the response of the channel remains roughly constant, that is, the cross-correlation is greater than one-half. In other words, the channel affects a range of frequencies $(\Delta f)_c$, from f_1 to f_2, similarly.

Therefore, narrowband signals that fit within the coherence bandwidth, experience nearly constant, or "flat," frequency fading. That implies that the

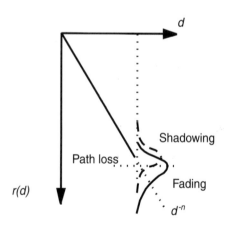

Figure 1.10 Shadowing and multipath propagation affect received signal strength.

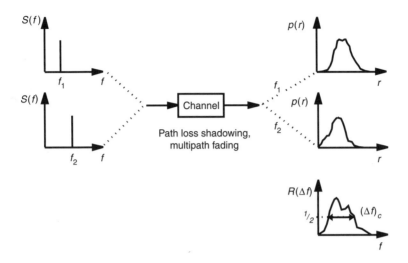

Figure 1.11 The frequency selective behavior of the channel affects the two transmitted signals differently.

transfer function of the communication channel is spectrally uniform, with constant gain and linear phase. Wideband signals, like the ones generated by direct-sequence spread-spectrum modulation,[2] typically extend beyond the coherence bandwidth and experience frequency-selective fading. With wideband signals, only a portion of the signal fades; thus, the integrity of the radio link is preserved through frequency diversity.

In the second example, two identical signals are transmitted at different times, t_1 and t_2, as shown in Figure 1.12. The channel affects each signal's received power level independently and produces distinct pdf's for the two output waveforms. The pdf's are cross-correlated to reveal changes in the channel. If $t_1 \approx t_2$, the cross-correlation of the two waveforms will be high. But as the separation between t_1 and t_2 increases, the cross-correlation will become lower and eventually fall below one-half. That indicates the time separation between signals where the channel response stays constant, that is, the time coherence of the channel, $(\Delta t)_c$. In other words, the response of the channel and the received power level is predictable as long as the separation in time between signals is less than the time coherence of the channel.

The coherence bandwidth and time coherence parameters are key measures of the communication channel. These parameters lead to a second set of parameters, known as the scattering functions, that describe the effect on

---

2. Most cellular CDMA systems, such as CDMA IS95 and WCDMA, use direct-sequence spread-spectrum modulation.

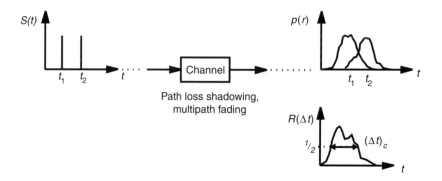

Figure 1.12 Time-varying behavior of the channel affects two pulses transmitted at separate times differently.

the transmitted signal. The scattering functions $S(\tau, \nu)$ are found by taking the Fourier transforms of the cross-correlation functions, that is,

$$R(\Delta f, \Delta t) \xleftrightarrow[\text{transform}]{\text{Fourier}} S(\tau, \nu) \qquad (1.3)$$

where the multipath delay spread, τ, is related to $1/(\Delta f)_c$ and the doppler spread, ν, is associated with $1/(\Delta t)_c$.

The cross-correlation parameters and scattering functions are small-scale effects caused by multipath propagation through the communication channel. These multipath rays are duplicate signals that are scaled and phase rotated relative to each other. Interestingly, at any instant t_o, the received signal is a composite of these replica signals. Consequently, the received signal at time t_o is described by

$$r^2(t_o) = \sum_{n=0}^{N-1} a_n^2(t_o) \qquad (1.4)$$

where a_n is the complex amplitude of the nth multipath rays.

The multipath delay spread (τ) is especially important in digital communication systems. It measures the smearing or spreading in the received signal when an impulse is transmitted through the communication channel. Impulse smearing is shown in Figure 1.13 for a typical cellular system. The first peak in the response generally corresponds to the line-of-sight ray, while the other peaks reveal the scaling and propagation delay of the strong multipath rays. The delay spread covers the time interval from the first peak to the last significant peak.

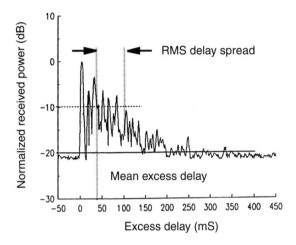

Figure 1.13 Measured multipath delay spread for a typical cellular system. (*From:* T. S. Rappaport, *Wireless Communications,* © 1995; reprinted by permission of Prentice-Hall, Inc., Upper Saddle River, NJ.)

The delay spread causes adjacent data bits to overlap and produces intersymbol interference (ISI). In narrowband communication systems, that can be disastrous and must be removed by equalization techniques. In wideband systems, it is possible to remove the delay from the multipath components and to align the rays using signal processing methods.[3] That yields the ensemble average of the received power,

$$E_{a,\theta}[P_r] = \sum_{n=0}^{N-1} a_n^2 \tag{1.5}$$

where the average is computed using all the multipath components. The striking result is that the aggregate power after alignment approaches the value due to lognormal shadowing, eliminating the multipath effects. Furthermore, in most situations, it is sufficient to consider only the largest multipath components, thereby simplifying the signal processing.

---

3. The most common approach to aligning the rays and constructively summing them is the Rake receiver, which is described in Chapter 5.

1.4.3 Modeling the Communication Channel

The wireless communication channel is unpredictable, making deterministic models of performance impossible [7–10]. As a result, the performance of wireless communication systems is assessed using simplifications of practical or particularly troublesome environments based on three basic models.

Figure 1.14(a) illustrates the simplest propagation model, line-of-sight propagation in a noisy environment. Here, the received signal is given by

$$r(t) = cs(t) + n(t) \tag{1.6}$$

where c is the path loss factor, $s(t)$ is the transmitted signal, and $n(t)$ is the added channel noise. The noise is constant over frequency and is usually referred to as white noise, while its amplitude is described by a zero-mean Gaussian pdf. The function is defined by

$$p(a) = \frac{1}{\sqrt{2\pi\sigma^2}} e^{-a^2/2\sigma^2} \tag{1.7}$$

where σ^2 is the variance of the random variable a. This type of noise source is called additive white Gaussian noise (AWGN). The line-of-sight model is appropriate for picocells or for wireline communications.

Wireless communication channels, however, are both time varying and frequency dependent. Therefore, the path loss factor of the line-of-sight model is altered to provide for the variation with time t and excess delay τ.[4] The received signal is then

$$r(t) = c(t, \tau) \cdot s(t) + n(t) \tag{1.8}$$

where $c(t, \tau)$ is a function that describes the wireless channel and models both large-scale and small-scale effects. By contrast, the line-of-sight model assumes that c is constant.

[əˈprɔksɪmɪt]

This second, improved model of the wireless channel is approximated in the following way. A signal $\cos\omega t$ is transmitted via the wireless channel and received at the receiver as $r\cos(\omega t + \phi)$, where r is a complex amplitude and ϕ is a uniformly distributed random variable. The complex amplitude r can be modeled as independent I and Q random variables [8]. Furthermore, there are a sufficient number of independent reflections (multipath rays) to allow those random variables to be modeled as Gaussian distributed with

---

4. The excess delay spread is tied to the coherence bandwidth $R(\Delta f)_c$.

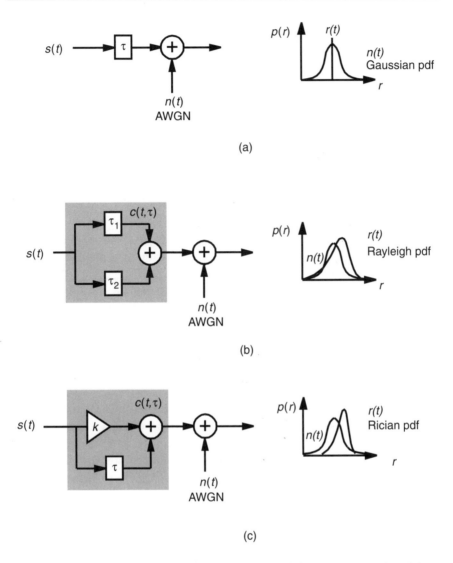

Figure 1.14 Channel models: (a) line of sight with AWGN, (b) Rayleigh channel model, and (c) Rician channel model.

$$r = \sqrt{I^2 + Q^2}.$$

The probability of receiving a signal of amplitude r follows a Rayleigh or Rician distribution that depends on the mean of the random variables I and Q. If the mean of both random variables is zero, the pdf of r is Rayleigh distributed and equal to

$$p(r) = \frac{r}{\sigma^2} e^{-(r^2/2\sigma^2)} \qquad (1.9)$$

where σ^2 is the time-averaged power level. That produces the channel model shown in Figure 1.14(b). If the mean of the random variables is nonzero, a dominant multipath component or a line-of-sight path is present and the pdf is Rician, that is,

$$p(r) = \frac{r}{\sigma^2} e^{-(r^2+A^2/2\sigma^2)} I_0\left(\frac{Ar}{\sigma^2}\right) \qquad (1.10)$$

where A is the peak of the dominant signal and $I_0(\cdot)$ is the modified Bessel function of the first kind and zero order. That leads to the channel model shown in Figure 1.14(c).

The Rician factor k describes the strength of the line-of-sight ray and equals

$$k = \frac{A^2}{2\sigma^2} \qquad (1.11)$$

As k approaches infinity, the Rician distribution becomes a delta function, which matches the simple line-of-sight model. As k approaches zero, the Rician distribution transforms into a Rayleigh distribution.

The AWGN, Rayleigh, and Rician channel models are simple, compact models for approximating the effects of radio propagation. An overview of more complicated models is available in [11].

1.5 Wireless Standards

It is vital to use the radio spectrum efficiently and to share the limited resource among multiple users. That requires multiple-access schemes that separate users by frequency, time, and/or orthogonal codes, as shown in Figure 1.15.

Most systems divide the radio spectrum into frequency channels and strategically assign those channels, a practice known as frequency division multiple access (FDMA). The channel assignment strategies minimize interference between users in different cells. Interference is caused by transmitted signals that extend outside the intended coverage area into neighboring cells. To limit interference, frequency channels are generally assigned based on the

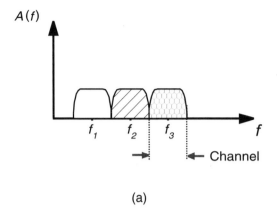

(a)

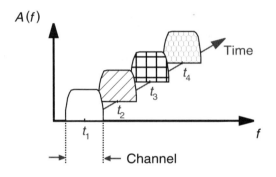

(b)

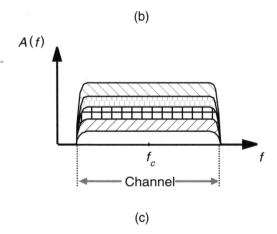

(c)

Figure 1.15 Multiple access methods: (a) frequency division multiple access (FDMA), (b) time division multiple access (TDMA), and (c) code division multiple access (CDMA).

frequency reuse pattern shown in Figure 1.16. In special cases, such as CDMA networks, universal frequency reuse is allowed and is a powerful advantage.

The choice of multiple-access technique directly affects subscriber capacity, which is a measure of the number of users that can be supported in a predefined bandwidth at any given time.

First-generation (1G) wireless communication systems use analog methods. These systems superimpose the message signal onto the RF carrier using frequency modulation (FM) and separate users by FDMA techniques. An example of this type of system is the Advanced Mobile Phone System (AMPS).

Second-generation (2G) communication systems introduce digital technology. These systems digitally encode the message signal before superimposing it onto the RF carrier. Digital data allows powerful coding techniques that both improve voice quality and increase network capacity. Examples of this type of system include GSM (Global System for Mobile Communications) [12], NADC (North American Digital Cellular) [13], PHS (Personal Handyphone System) [14], and CDMA IS95 [15].

Table 1.2 compares some of the leading wireless standards.

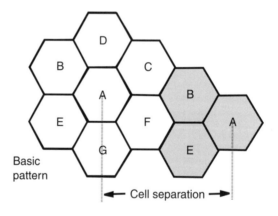

Figure 1.16 Seven-cell reuse pattern typically used by carriers to separate frequency channels.

Table 1.2
Important Properties of Some Leading Wireless Standards

Standard	AMPS	GSM	NADC	PHS	CDMA IS95
Frequency plan					
Tx (MHz)	824–849	880–915	824–849		824–849
Rx (MHz)	869–894	925–960	869–894		869–894
Tx (MHz)		1,710–1,785	1,850–1,910	1,895–1,907	1,850–1,910
Rx (MHz)		1,805–1,885	1,930–1,990	1,895–1,907	1,930–1,990
Multiple access	FDMA	F/TDMA	F/TDMA	F/TDMA	F/CDMA
Channel spacing (Hz)	30K	200K	30K	300K	1.25M
Modulation	FM	GMSK	π/4QPSK	π/4DQPSK	QPSK
Maximum Tx power		1W	600mW	80mW	200mW
Bit rate	NA	13 Kbps	8 Kbps	32 Kbps	1–8 Kbps
Speech per channel	1	8	3	4	28
Number of users (in 10 MHz spectrum)	47	56	142	19	224

References

[1]   Dataquest Survey of Worldwide Wireless Subscribers, Nov. 1999.

[2]   Modarressi, A. R., and R. A. Skoog, "Signaling System No. 7: A Tutorial," *IEEE Communications Magazine*, July 1990, pp. 19–35.

[3]   Russel, T., *Signaling System #7*, New York: McGraw-Hill, 1998.

[4]   Gallagher, M. D., and R. A Snyder, *Mobile Telecommunications Networking*, New York: McGraw-Hill, 1997.

[5]   Stallings, W., *Handbook of Computer Communications Standards—The Open Systems Interconnection (OSI) Model and OSI-Related Standards*, New York: Macmillan, 1987.

[6]   Stremler, F. G., *Introduction to Communication Systems*, Reading, MA: Addison-Wesley, 1992.

[7]   Rappaport, T. S., *Wireless Communications: Principles and Practice*, Upper Saddle River, NJ: Prentice Hall, 1996.

[8]   Steele, R., ed., *Mobile Radio Communications*, New York: IEEE Press, 1992.

[9]   Proakis, J. G., *Digital Communications*, New York: McGraw-Hill, 1995.

[10] Anderson, J. B., and T. S. Rappaport, "Propagation Measurements and Models for Wireless Communications Channels," *IEEE Communications Magazine*, Jan. 1995, pp. 42–49.

[11] Adawi, N. S., et al., "Coverage Prediction for Mobile Radio Systems Operating in the 800/900 MHz Frequency Range," *IEEE Trans. on Vehicular Technology*, Vol. 37, No. 1, Feb. 1988.

[12] Mouly, M., and M. B. Pautet, *The GSM System for Mobile Communications*, 1992.

[13]   TIA/EIA Interim Standard, "Cellular System Dual Mode Mobile Station-Base Station
        Compatibility Standard," IS-54B, Apr. 1992.

[14]   Personal Handiphone System RCR Standard 28, Ver. 1, Dec. 20, 1993.

[15]   TIA/EIA Interim Standard, "Mobile Station-Base Station Compatibility Standard for
        Dual-Mode Wideband Spread Spectrum Cellular System," IS-95A, Apr. 1996.

2

The CDMA Concept

CDMA is a multiple-access scheme based on spread-spectrum communication techniques [1–3]. It spreads the message signal to a relatively wide bandwidth by using a unique code that reduces interference, enhances system processing, and differentiates users. CDMA does not require frequency or time division for multiple access; thus, it improves the capacity of the communication system.

This chapter introduces spread-spectrum modulation and CDMA concepts. It presents several design considerations tied to those concepts, including the structure of the spreading signal, the method for timing synchronization, and the requirements for power control. This chapter also points out CDMA IS95 [4] details to illustrate practical solutions to these design issues.

2.1 Direct-Sequence Spread-Spectrum Communications

Spread-spectrum communications is a secondary modulation technique. In a typical spread-spectrum communication system, the message signal is first modulated by traditional amplitude, frequency, or phase techniques. A pseudo-random noise (PN) signal is then applied to spread the modulated waveform over a relatively wide bandwidth. The PN signal can amplitude modulate the message waveform to generate direct-sequence spreading, or it can shift the carrier frequency of the message signal to produce frequency-hopped spreading, as shown in Figure 2.1.

The direct-sequence spread-spectrum signal is generated by multiplying the message signal $d(t)$ by a pseudorandom noise signal $pn(t)$:

$$g(t) = pn(t)d(t) \qquad (2.1)$$

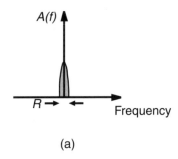

(a)

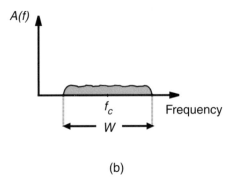

(b)

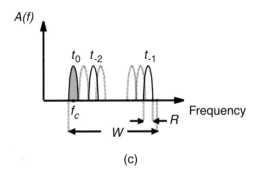

(c)

Figure 2.1 Spread-spectrum signals: (a) message signal, (b) direct-sequence signal, and (c) frequency-hopped signal.

In most cases, the PN signal is a very high rate, nonreturn-to-zero (NRZ) pseudorandom sequence that chops the modulated message waveform into chips, as shown in Figure 2.2. Hence, the rate of the secondary modulating waveform is called the chip rate, f_c, while the rate of the message signal is designated the bit rate, f_b. The two modulation processes produce different bandwidths, namely, R for the modulated message signal and W for the relatively wide spread-spectrum waveform. Note that the secondary modulation does not increase the overall power of the message signal but merely spreads it over a wider bandwidth.

The frequency-hopped spread-spectrum signal is formed by multiplying the message signal with a pseudorandom carrier frequency $\omega_{pn}(t)$:

$$g(t) = \cos[\omega_{pn}(t)t]d(t) \qquad (2.2)$$

In this approach, the spectrum of the modulated message hops about a range of frequencies and produces a relatively wide bandwidth signal.

Spread-spectrum modulation techniques provide powerful advantages to communication systems, such as a flexible multiple-access method and interference suppression. These advantages are examined here for direct-sequence spread-spectrum signals.

The direct-sequence spread-spectrum signal formed in a simple and ideal transmitter can be described by

$$s(t) = pn(t)Ad(t)\cos(\omega t + \theta) \qquad (2.3)$$

where $pn(t)$ is the pseudorandom modulating waveform, A is the amplitude of the message waveform, $d(t)$ is the message signal with bipolar values ± 1,

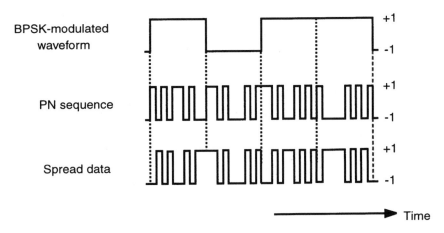

Figure 2.2 Direct-sequence spread-spectrum signals.

ω is the carrier frequency, and θ is a random phase. The signal is transmitted over the air interface and is received along with thermal noise $n(t)$ and interference $i(t)$, which are added by the channel. The received signal is[1]

$$r(t) = pn(t)Ad(t)\cos(\omega t + \theta) + n(t) + i(t) \tag{2.4}$$

To recover the message signal $d(t)$, the RF carrier, $\cos(\omega t + \theta)$, is removed, and the spread-spectrum signal is despread by a simple correlator. The correlator is synchronized to the transmitter's sequence, $pn(t)$, and its output is integrated over the bit period (T_b). The process is described by

$$\frac{1}{T_b}\int_0^{T_b} pn(t)r(t)dt = pn^2(t)Ad(t) + pn(t)[n'(t) + i'(t)] \approx Ad(t) \tag{2.5}$$

where $n'(t)$ and $i'(t)$ represent the down-converted thermal noise and interference. When the PN sequences at the transmitter and the receiver are synchronized, $pn^2(t) = 1$ and the bit energy is compressed back to its original bandwidth R. Any received interference, $i(t)$, is spread by the correlator to the relatively wide bandwidth W, and its effect is lowered.

The correlator affects the message signal $d(t)$ differently than it does the interference $i(t)$ and thereby improves the signal-to-noise ratio (SNR) of the received signal.[2] That powerful benefit is the *processing gain* of the system and is equal to the spreading factor W/R.

2.1.1 Spreading Codes

The spreading code is a critical component of spread-spectrum communications. It generates the pseudorandom signal used to spread the message signal. To be effective, the spreading code should produce values that resemble Gaussian noise and approximate a Gaussian random variable. In addition, these codes should be easily realizable at the transmitter and the receiver.

In general, the spreading signal is a binary waveform with values specified at the chip rate. The binary waveform allows easy implementation without sacrificing performance and enables synchronization of the transmitter to the received signal. It is possible to achieve a continuous-time waveform by passing the binary signal through a linear filter.

---

1. To illustrate the spread-spectrum concept, delay and scaling effects introduced by the channel are ignored here.
2. *Noise* refers to any unwanted energy and includes interference.

These characteristics are available from deterministic, pseudorandom sequences with the following classical properties:

- There are near-equal occurrences of +1 and −1 chips.
- Run lengths of r chips with the same sign occur approximately 2^{-r} times.
- Shifting by a nonzero number of chips produces a new sequence that has an equal number of agreements and disagreements with the original sequence [1].

The randomness of the signal $pn(t)$ is measured by the autocorrelation function $R_{pn}(\tau)$, given by

$$R_{pn}(\tau) = \lim_{T \to \infty} \frac{1}{T} \int_{-T/2}^{T/2} pn(t)pn(t + \tau)dt \qquad (2.6)$$

Similarly, the autocorrelation for a sequence of M discrete values is written as

$$R_{pn}(\tau) = \frac{1}{M} \sum_{M} pn(t)pn(t + \tau) \qquad (2.7)$$

and is plotted in Figure 2.3. A peak or peaks in the function indicate that the sequence contains subsequences that repeat. For a properly designed PN sequence, the autocorrelation function is very small and equal to $-1/M$ for every nonzero value of τ. Consequently, PN sequences also are useful for timing synchronization.

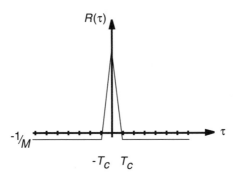

Figure 2.3 Autocorrelation of PN sequence.

The uniqueness of the signal $pn(t)$ is analyzed with the cross-correlation function, defined by

$$R_{xy}(\tau) = \lim_{T\to\infty} \frac{1}{T} \int_{-T/2}^{T/2} x(t)y(t + \tau)dt \tag{2.8}$$

or, alternatively, by

$$R_{xy}(\tau) = \frac{1}{M} \sum_{M} x(t)y(t + \tau) \tag{2.9}$$

where $x(t)$ and $y(t)$ are two different signals or sequences. In general, pseudo-random sequences demonstrate poorer cross-correlation attributes than deterministic sequences, as shown in Figure 2.4 [5]. That is because orthogonal sequences are designed to be dissimilar or orthogonal to each other. As a result, orthogonal codes are used in CDMA systems to differentiate users and to minimize interference.

The Hadamard code is a commonly used orthogonal code [6]. It is based on the rows of a square (n by n) matrix known as the Hadamard matrix. In the matrix, the first row consists of all 0s, while the remaining rows contain equal occurrences of 0s and 1s. Furthermore, each code differs from every other code in $n/2$ places.

The Hadamard matrix if formed by the following recursive procedure:

$$W_1 = [0] \quad W_2 = \begin{bmatrix} W_1 & W_1 \\ W_1 & \overline{W_1} \end{bmatrix} = \begin{bmatrix} 0 & 0 \\ 0 & 1 \end{bmatrix} \quad W_{2n} = \begin{bmatrix} W_n & W_n \\ W_n & \overline{W_n} \end{bmatrix} \tag{2.10}$$

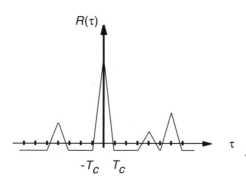

Figure 2.4 Cross-correlation of a PN sequence.

where $\overline{W}_n$ is derived from W_n by replacing all entries with their complements. The Hadamard matrix provides n orthogonal codes.

2.1.2 Spread-Spectrum Performance

Spread-spectrum modulation and CDMA techniques allow several users to share the radio interface; thus, the received waveform becomes the sum of k user signals and noise:

$$r(t) = \sum_{n=1}^{k} pn_n(t)A_n d_n(t)\cos(\omega t + \theta_n) + n(t) \tag{2.11}$$

The receiver retrieves the message signal by despreading the received signal. It does that by synchronizing its correlator to a specific spreading sequence, $pn(t)$, that is unique to the user and different from those of other users. As a result, the other user signals appear noiselike.

The noise (N_t) seen by the correlator is the signal energy received from the $k - 1$ users and thermal noise, that is,

$$N_t = \sum_{n=1}^{k-1} S_n + WN_o \tag{2.12}$$

where S_n is the received power from the nth user, N_o is the thermal noise power spectral density (psd), and W is the channel bandwidth. If the received power from each user is assumed equal[3] and k is large, such that $k - 1$ can be approximated by k, then

$$N_t \approx kS + WN_o \tag{2.13}$$

Furthermore, the interference generally is much larger than the integrated thermal noise ($kS \gg WN_o$), so that

$$N_t \approx I = kS \tag{2.14}$$

From this result, two important observations are made. First, the interference of the spread-spectrum system increases linearly as the number of users is added. Second, the performance of the system suffers when any user transmits extra power, a problem known as the near-far effect [3].

---

3. This assumption is valid because in CDMA systems the power received from each user is strictly controlled.

The SNR is a key consideration in all communication systems. In digital communication systems, the SNR is characterized by a related figure of merit, the bit energy per noise density ratio (E_b/N_o). That parameter takes into account the processing gain of the communication system, a vital consideration in spread-spectrum communications. The parameter normalizes the desired signal power to the bit rate R to determine the bit energy and the noise or interference signal power to the spreading bandwidth W to determine the noise spectral density. Recall that the correlator

- Despreads or integrates the desired signal to the narrow bandwidth of the original message signal (R);
- Spreads the interference to a wider bandwidth;
- Leaves the uncorrelated noise unaltered.

Therefore,

$$\frac{E_b}{N_o} = \frac{S/R}{N/W} \approx \frac{S}{kS(R/W)} \tag{2.15}$$

Amazingly, the interference from other users (i.e., self-interface) is reduced by the processing gain (W/R) of the system.

A simple expression for the capacity of a CDMA system is developed from (2.15) and is given by

$$k \approx \frac{W/R}{(E_b/N_o)_{min}} \tag{2.16}$$

where $(E_b/N_o)_{min}$ is the minimum value needed to achieve an acceptable level of receiver performance, typically measured as the bit error rate (BER). The expression shows that the capacity of CDMA communication systems depends heavily on the spreading factor and the receiver's performance. The capacity is tied to a flexible resource—power—and is said to be *soft-limited.* In other words, if the required E_b/N_o is lowered, the transmit signal power allocated to each user is reduced, and the number of users can be increased. In contrast, the capacity of systems that employ other multiple-access methods like FDMA and TDMA are hard-limited. That is because their capacity is fixed by system design.

2.2 Overview of the CDMA IS95 Air Interface

Spread-spectrum communications using CDMA techniques originally were developed for military use [7]. The systems provided vital anti-jamming and low probability of intercept (undetectable) properties. Later, it was realized that those techniques also benefited cellular communications over dispersive channels. That led to operational (CDMA IS95) and planned (next-generation CDMA) networks based on spread-spectrum communications.

CDMA IS95 is a recent 2G wireless protocol. It and other 2G wireless protocols provide increased capacity, more robust service, and better voice quality by introducing digital methods. The CDMA IS95 standard [4] describes implementation details of the network, including the air interface, the protocol stack, the base station and mobile radio transmitters, the spreading codes, and the power control requirements.

2.2.1 Forward Link

The base station transmits radio signals to the mobile radio and forms the forward link, or downlink.[4] It relies on the forward-link modulator to protect the message signal against radio propagation impairments, to perform spread-spectrum modulation, and to provide multiple access by code division. The forward-link modulator is shown in Figure 2.5, and its operation is outlined below.

The input to the modulator is digital data from the voice coder (vocoder) or an application. The signal is protected using a forward error correction code (convolutional code) and repeated as needed to fill the frame buffer. Each frame buffer is then time interleaved to protect against burst errors. The time-interleaved data stream is scrambled by the long PN sequence, which has been slowed to match the bit rate. Power control information is then added. The resulting data is spread using an orthogonal Walsh code and randomized by the short in-phase (I) and quadrature-phase (Q) PN sequences. The signal then is applied to an RF carrier and transmitted.

The interleaving process scatters the bit order of each frame so that if a segment of data is lost during fading, its bits are dispersed throughout the reorganized frame, as illustrated in Figure 2.6. The missing bits are often recovered during the decoding process. Interleaving provides effective protection against rapidly changing channels but hinders performance in slow-changing environments.

The long code provides privacy by scrambling the message data. The short PN sequences distribute the energy of the transmit waveform so it appears

4. The term *downlink* is a carryover from satellite communications.

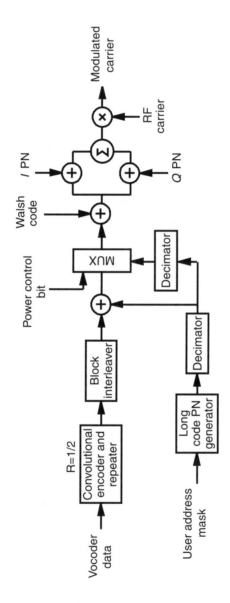

Figure 2.5 Forward-link modulator for CDMA IS95 base station.

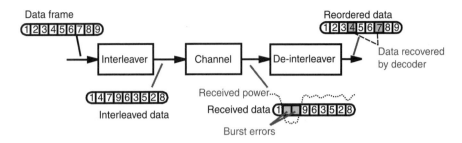

Figure 2.6 Interleaving process provides protection against time-varying channels.

Gaussian and noiselike. Neither of those PN codes spreads the message signal to the wide spread-spectrum bandwidth. It is the Walsh code that provides the orthogonal spreading. It multiplies each message signal by a 64-bit Walsh code unique to each user and spreads the signal bandwidth. As a result, a 64x processing gain is obtained.

The forward link contains several logical channels: the pilot channel, the synchronization (sync) channel, up to seven paging channels, and, at most, 55 traffic channels. The pilot, sync, and paging channels are common control channels, shared by all the users in the cell coverage area, which support communications between the mobile radios and the base station. The traffic channels are dedicated channels that support user communications. The channels are assigned to unique Walsh codes, as shown in Figure 2.7, and are able to share the air interface with very little interaction.

The pilot channel serves three purposes: channel estimation for coherent demodulation, multipath detection by the receiver, and cell acquisition during handoff (a procedure that maintains the radio link as a mobile radio moves from one cell coverage area to another). The pilot channel is a common channel that is broadcast to multiple users. As such, the overhead of the channel is divided by the number of users in the cell coverage area. That means it can be allocated more energy to improve performance without significant impact.

The pilot channel uses Walsh channel 0 (the all-zero entry in the Hadamard matrix) and an all-zero data sequence. Therefore, the pilot channel is just a replica of the short PN sequences. Because the pilot channel is a PN

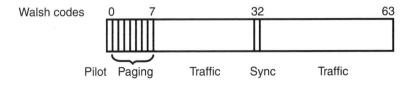

Figure 2.7 Forward-link channels in CDMA IS95 systems.

sequence, it displays good autocorrelation properties and provides a means for timing synchronization, an important aspect of the CDMA IS95 network.

The short PN sequence is a sequence of 2^{15} chips that is conveniently written about a circle, as shown in Figure 2.8. The figure illustrates the periodicity and pseudorandom characteristics of the PN sequence. The short PN sequence is divided into consecutive segments that are 64 chips long, and each segment is labeled with an offset value[5] relative to the top of the circle. The base stations in the network are assigned to different offsets and are therefore synchronous to each other.

Neighboring base stations are typically separated by 12 PN offsets, equal to 625 μs. By comparison, typical values for multipath delay spread lie between a few hundred nanoseconds and a few microseconds. As a result, pilot signals from neighboring base stations are clearly distinguishable from any multipath rays.

The sync channel is assigned to Walsh code 32 and used for system timing. The base station transmits several messages on that channel at a data rate of 1.2 Kbps. One of the messages is the pilot PN offset, which is a reference point for the short PN sequence. Another message is the value of the long-code generator advanced by 320 ms. That is used to offset or rotate the mobile's PN generator and align it to the base station. In CDMA IS95, the base stations rely on the global positioning system (GPS) for system timing and to establish a synchronous network. The following messages also are transmitted by way of the sync channel: the communication air interface (CAI) reference level,

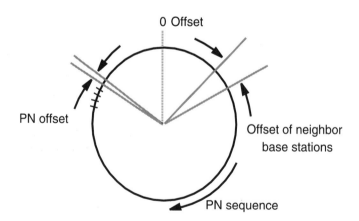

Figure 2.8   Short PN sequence written in circular form to show pattern and rotation of PN offsets.

---

5. The offsets indicate a rotation in time of a common PN sequence.

the system identification (SID) number, and the paging channel data rate (9.6, 4.8, or 2.4 Kbps).

The paging channel is used to control the base station to mobile link and is assigned to one of seven Walsh channels (codes 1–7). The base station uses this channel to wake up the mobile, respond to access messages, relay overhead information, and support handoff functions. It communicates several overhead messages, including the neighbor list. The neighbor list contains the PN offsets of nearby base stations, which accelerates pilot acquisition during handoff in a synchronous network. The paging channel also assigns the subscriber to one of the available traffic channels.

Traffic channels are assigned to the remaining 55 Walsh codes. These channels carry information at one of two primary rates: 8 Kbps (rate set 1) and 13 Kbps (rate set 2).[6] It is possible to lower the voice data rate during low speech activity periods, such as pauses that occur during listening, by using a variable rate vocoder (Chapter 4 covers speech coding) [8]. These algorithms support full, half, quarter, and one-eighth data rates that reduce system interference.

Table 2.1 summarizes the data rates and the channel coding characteristics of forward-link channels.

The message data are divided into blocks known as frames. Each frame consists of 192 symbols and spans 20 ms. This is a convenient period because speech signals appear pseudostationary over short periods of time, typically 5 to 20 ms, while longer periods of time produce noticeable distortion to the listener. Each 20-ms block of speech is analyzed to determine its content and to set the vocoder rate.

Each speech frame is appended with CRC and tail bits, as shown in Figure 2.9. The CRC is a parity check that is available at most data rates[7] and

Table 2.1
Forward-Link Channel Parameters for CDMA IS95 System

Channel	Data Rate (Kbps)	Channel Coding	Access Method	Processing Gain
Pilot	—	None	Walsh 0	—
Sync	1.2	Rate 1/2	Walsh 32	1024
Paging	4.8, 9.6	Rate 1/2	Walsh 1-7	128, 256
Traffic				
Rate set 1	$1.2N$	Rate 1/2	Walsh 8-31, 33-63	$1024/N$
Rate set 2	$1.8N$	Rate 1/2		$682.6/N$

6. The rates in Table 2.1 are higher because these include parity bits.
7. The CRC is available at full and half rates for rate set 1 and all rates for rate set 2.

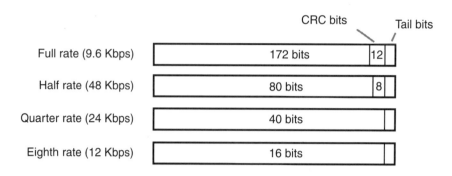

Figure 2.9 Forward-link frame structure in CDMA IS95 systems.

is used to assist rate determination. The tail bits are used to flush out the convolutional encoder after each frame is processed.

The variable rate vocoder increases the capacity of the CDMA IS95 communication system. That is because at half-rate, each symbol is transmitted twice at one-half the nominal power; at quarter-rate, each symbol is transmitted four times at one-fourth the nominal power; and at eighth rate, each symbol is transmitted eight times at one-eighth the nominal power. That achieves the same energy per bit at the receiver but progressively lowers the transmit power.

Another way to increase capacity in a communication system is to limit the transmit energy outside the channel bandwidth. The base station transmitter includes a bandwidth-shaping filter for that purpose. It is a Chebyshev equiripple finite impulse response (FIR) filter with an extremely narrow transition band. The transmitter also includes an all-pass filter to compensate for group delay distortion expected at the mobile radio receiver. Group delay and phase distortion are critical parameters for phase-modulated communication systems.

Table 2.2 lists the minimum performance requirements for a cellular-band mobile radio receiver.[8] For these tests, the connecting base station transmits a full suite of channels at defined power levels. The CDMA IS95 standard does not provide any additional information regarding the mobile radio receiver. Its design is proprietary to each manufacturer and is extremely challenging.

2.2.2 Reverse Link

The mobile radio transmits signals to the base station and thereby forms the reverse link, or uplink. It employs the reverse-link modulator to protect the message signal against radio propagation impairments and to align to system

---

8. The minimum performance requirements specify the power levels assigned to the pilot, sync, paging, and interfering users as well as the desired user.

Table 2.2
Minimum Performance Requirements for CDMA IS95 Mobile Radio Receiver

Parameter	Conditions	Requirement
Sensitivity	FER < 0.005	−104 dBm
Maximum input	FER < 0.005	−25 dBm
Single tone desensitization	Adjacent channel @ −30 dBm FER < 0.01	−101 dBm
Low-level intermodulation distortion (IMD)	Adjacent channel @ −40 dBm FER < 0.01	−101 dBm
High-level IMD	Adjacent channel @ −21 dBm Alternate channel @ −21 dBm FER < 0.01	−79 dBm

timing. The reverse-link modulator is shown in Figure 2.10 and its operation is outlined next.

Unlike the forward link, it is nearly impossible to establish truly orthogonal traffic channels on the reverse link. That is because the mobile radios are located randomly in the cell area, at different distances to the base station, and with different propagation delays. As such, synchronization breaks down and spreading codes become less effective. Mobile radios are further constrained by portable operation and other consumer form-factor requirements. Consequently, the reverse-link modulator is comparatively simple, and the performance burden of the reverse link is shouldered by the base station.

The input to the reverse-link modulator is digital data from the vocoder or an application. The signal is encoded and repeated to fill the frame buffer. The data is then interleaved and Walsh-modulated. Each frame is then divided into 16 equal sets of data called power control groups. When the vocoder is running at less than full-rate, the repeater and the interleaver work together to produce duplicate sets of data within the frame. The details are fed forward to control the data burst randomizer, which pseudorandomly blanks redundant data. The transmitter is punctured off (turned off) during blank periods, thereby lowering its time-averaged output power. The resulting data stream is then multiplied by the masked long code and randomized by the I and Q channels short PN codes. Both the short and long codes are synchronized to the base station using information received on the sync channel.

Walsh modulation is a 64-ary modulation method that translates 6-bit symbols to one of 64 modulation states. Each modulation state is a 64-bit entry from the 64-by-64 Hadamard matrix used by the forward-link modulator. The difference is that here the Hadamard matrix is used to define the distinct

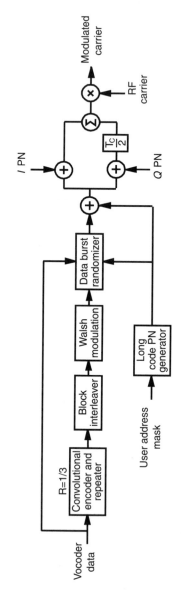

Figure 2.10 Reverse-link modulator for CDMA IS95 mobile radio.

points (or modulation states) of the constellation and is not used for spreading or multiple access.

The reverse link contains two types of channels, as shown in Figure 2.11. The access channel is the complement to the forward link's paging channel. It is used to originate calls, respond to pages, register the mobile phone, and communicate other overhead messages. It transmits data at 4.8 Kbps. The other type of channel is the traffic channel, which carries the message signal and uses the Walsh code assigned by the base station.

The long code, which is masked by the electronic serial number (ESN) of the mobile, is used to distinguish between CDMA users on the reverse link. (The masking operation is described in Section 5.1.2.) It provides pseudo-orthogonal PN spreading of the users on the reverse link based on its autocorrelation properties. There are up to 32 access channels (for each dedicated paging channel) and as many as 62 traffic channels on the reverse link. In practice, fewer traffic channels are allowed because of minimum performance requirements.

Table 2.3 summarizes the data rate and channel coding characteristics of reverse-link channels. Table 2.4 lists the minimum performance requirements for the mobile radio transmitter. The requirements ensure the quality of the reverse link and help maximize network capacity.

The waveform quality factor (ρ) measures the modulation accuracy using the cross-correlation of the transmitted signal to the ideal baseband signal [9], that is,

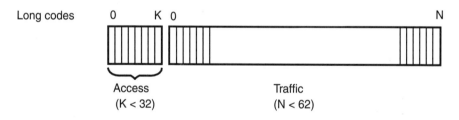

Figure 2.11 Allocation of reverse-link channels in CDMA IS95 systems.

Table 2.3
Reverse-Link Channel Parameters for CDMA IS95 Systems

Channel	Data Rate (Kbps)	Channel Coding	Access Method	Processing Gain
Access	4.8	Rate 1/3	Long-code mask	4
Traffic				
Rate set 1	$1.2N$	Rate 1/3	Long-code mask	4
Rate set 2	$1.8N$	Rate 1/2	Long-code mask	4

Table 2.4
Minimum Performance Requirements for IS95 CDMA Mobile Radio Transmitter

Parameter	Conditions	Capability
Maximum RF level		+23 dBm
Minimum controlled RF level	−50 dBm	
Adjacent channel power	900 kHz offset	−42 dBc/30 kHz
	2.385 MHz offset	−55 dBm/1 MHz
Alternate channel power	1.98 MHz offset	−54 dBc/30 kHz
	2.465 MHz offset	−55 dBm/1 MHz
Waveform quality		$\rho > 0.944$

$$\rho = \frac{\sum_{k=1}^{M} D_k S_k}{\sum_{k=1}^{M} D_k \sum_{k=1}^{M} S_k} \tag{2.17}$$

where S_k is the kth sample of the transmitted signal, D_k is the kth sample of the ideal baseband signal, and M is the measurement period in half-chip intervals. In practice, the waveform quality factor usually measures about or above 0.98 [10].

2.2.3 Power Control Algorithm

The user capacity in direct-sequence CDMA is limited by self-interference and adversely affected by the near-far problem at the base station receiver. Thus, accurate power control of all the mobile radio transmitters in the system is essential and an added challenge for the transceiver design. The receiver includes an automatic gain control (AGC) loop to track the received power level, which varies because of large-scale path loss and small-scale fading. To compensate for those effects, CDMA IS95 employs two power control methods.

The open-loop method uses the power level at the mobile radio receiver (P_{Rx}) to estimate the forward-link path loss. It then specifies the transmit power (P_{Tx}) of the mobile radio as

$$P_{Tx} \approx -73dBm - P_{Rx} \tag{2.18}$$

For example, if the received power level is −85 dBm, then the transmit power level is adjusted to +12 dBm. Note that the response of the open-loop method is made intentionally slow, as shown in Figure 2.12, to ignore small-scale fading.

Adding a feedback signal completes the AGC loop and improves the accuracy of the open-loop method. The feedback signal is an error signal sent from the base station to the mobile radio that instructs the mobile radio to increase or decrease power by a set amount, generally 1 dB. It is sent once per power control group and is therefore updated at a rate of 800 Hz. As such, it is sufficient to support vehicle speeds up to 100 km/h [11]. This second power control method is referred to as closed-loop power control.

2.2.4 Performance Summary

Communication systems are designed to provide high quality services to as many subscribers as possible. The tradeoff between the maximum number of subscribers and the quality of service is not straightforward in CDMA communication networks.

In direct-sequence spread-spectrum CDMA systems, capacity is soft limited by self-interference. The interference in this system was given in (2.14) as $I \approx kS$. In CDMA IS95 systems, that interference is reduced by the lower transmit power due to the variable rate vocoder and is increased by adjacent cells using the same frequency channel. As a result,

$$I \approx kS(1 + f)\nu \tag{2.19}$$

where f is a factor that accounts for "other-cell" interference effects (on average 0.55) [12] and ν is the voice activity rate (typically 3/8 for English speech) [13].

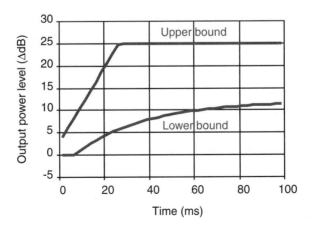

Figure 2.12 Open-loop power response of the mobile radio transmitter.

In practice, where high capacity is needed, each cell is sectored using directional antennas. For a three-sector cell, that provides an antenna gain (G_s) of about 2.5 [14]. Consequently, the capacity of the reverse link of a CDMA IS95 cell is

$$k \approx \frac{G_s}{\nu(1 + f)} \frac{W/R}{(E_b/N_o)_{min}} \qquad (2.20)$$

where ideal power control is assumed. The minimum value of E_b/N_o depends on the communication channel and the required performance of the receiver. For low mobility, the channel becomes more predictable, power control methods improve, while interleaving breaks down. In that situation, $(E_b/N_o)_{min}$ is about 4 dB and the estimated capacity is 46 users/cell. For high mobility, interleaving performs well but power control falls apart. There, the required E_b/N_o is approximately 6 dB and the estimated capacity is 29 users/cell. Of course, those numbers will be lower with nonideal power control [9].

The forward link is limited differently. Power control within the cell is ideal because all the transmit signals originate from a single base station and experience similar radio propagation effects. (Power control still is needed to minimize cell-to-cell interference.) In CDMA IS95 systems, the forward link is actually limited by available Walsh codes and soft handoff effects. To improve performance through spatial diversity and to assist handoff, a mobile user usually is linked to more than one base station, a situation known as soft handoff. Each connection requires a dedicated traffic channel and Walsh code. In fact, field tests show each user occupies, on average, 1.92 traffic channels. Therefore, the capacity of the forward link is

$$k \approx \frac{m}{1.92} \qquad (2.21)$$

where m is the number of Walsh codes. Since $m = 55$, the capacity is 28, which is lower than the reverse link. Surprisingly, the user capacity of CDMA IS95 is limited by the forward link, even though the reverse-link channels are not orthogonal.

References

[1]   Pickholtz, R. L., D. L. Schilling, and L. B. Milstein, "Theory of Spread-Spectrum Communications—A Tutorial," *IEEE Trans. on Communications*, Vol. 30, No. 5, May 1982, pp. 855–884.

[2]   Peterson, R. L., R. E. Ziemer, and D. E. Borth, *Introduction to Spread Spectrum Communications*, Upper Saddle River, NJ: Prentice Hall, 1995.

[3]   Cooper, G. R., and C. D. McGillen, *Modern Communications and Spread Spectrum*, New York: McGraw-Hill, 1986.

[4]   TIA/EIA Interim Standard, "Mobile Station-Base Station Compatibility Standard for Dual-Mode Wideband Spread Spectrum Cellular System," IS95a, Apr. 1996.

[5]   Simon, M. K., et al., *Spread Spectrum Communications Handbook*, New York: McGraw-Hill, 1994.

[6]   Rappaport, T. S., *Wireless Communications: Principles and Practice*, Upper Saddle River, NJ: Prentice Hall, 1996.

[7]   Pickholtz, R. L., L. B. Milstein, and D. L. Schilling, "Spread Spectrum for Mobile Communications," *IEEE Trans. on Vehicular Technology*, Vol. 40, No. 2, May 1991, pp. 313–322.

[8]   Padovani, R., "Reverse Link Performance of IS95 Based Cellular Systems," *IEEE Personal Communications*, Third Quarter 1994, pp. 28–34.

[9]   Birgenheier, R. A., "Overview of Code-Domain Power, Timing, and Phase Measurements," *Hewlett-Packard J.*, Feb. 1996, pp. 73–93.

[10] Chen, S.-W., "Linearity Requirements for Digital Wireless Communications," *IEEE GaAs IC Symp.*, Oct. 1997, pp. 29–32.

[11] Salmasi, A., and K. S. Gilhousen, "On the System Design Aspects of Code Division Multiple Access (CDMA) Applied to Digital Cellular and Personal Communication Networks," *Proc. IEEE Vehicular Technology Conf.*, VTC-91, May 1991, pp. 57–63.

[12] Viterbi, A. J., et al., "Other-Cell Interference in Cellular Power-Controlled CDMA," *IEEE Trans. on Communications*, Vol. 42, No. 4, pp. 1501–1504, Apr. 1994.

[13] Brady, P. T., "A Statistical Analysis of On-Off Patterns in 16 Conversations," *Bell Systems Tech. J.*, Vol. 47, Jan. 1968, pp. 73–91.

[14] Garg, V. K., K. Smolik, and J. E. Wilkes, *Applications of CDMA in Wireless/Personal Communications*, Upper Saddle River, NJ: Prentice Hall, 1997.

3

The Digital System

Modern communication systems increasingly rely on the digital system for sophisticated operations and advanced signal processing routines. Typical mobile radio architectures include two specialized computers: the MCU, which supervises management functions, and the DSP, which executes key signal processing algorithms.

More and more signal processing is being performed digitally because of developments in complementary metal oxide semiconductor (CMOS) technology and improvements in DSP architecture. CMOS very large scale integration (VLSI) technology offers low-power, low-cost, and highly integrated solutions that continue to shrink. Amazingly, CMOS transistor density continues to double every eighteen months [1]. DSP architecture improvements make possible the powerful algorithms that are vitally needed to enhance the performance of wireless communication systems.

Signal processing functions are implemented in firmware and specialized hardware. Firmware designs provide flexibility but typically consume more power; in contrast, hardware designs generally run faster and consume less power.

This chapter describes the general-purpose MCU and the application-specific DSP. It covers some of the management tasks handled by the MCU, including protocol administration and power management. It concludes with fundamental digital signal processing operations, such as sampling, sample rate conversion, digital filtering, spectral analysis, data windowing, and data detection.

3.1 Architecture Issues

The digital system consists of an MCU, at least one DSP, and extensive memory, as shown in Figure 3.1. It typically uses two bus sets,[1] one set for instructions and the other for data, to keep the processors fed and to reduce computation times. It also provides the user interface (display, keypad, microphone, and speaker), connects to the RF transceiver, and supports external communications.

3.1.1 The MCU

The MCU supervises the operation of the mobile radio and administers the procedures associated with the communication protocol. It relies on a state-of-the-art microprocessor and includes an arithmetic logic unit (ALU), timers, and register files.

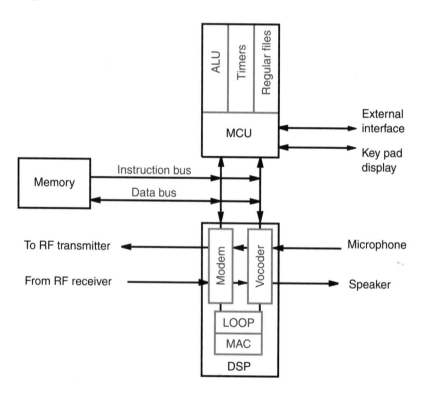

Figure 3.1 The digital system in a modern mobile radio.

---

1. A bus set includes an address bus and a data bus.

The ALU performs various logic functions, such as comparisons, and supports computationally demanding tasks. It may also incorporate specialized hardware to accelerate division, square root, and other special math functions.

The MCU assigns timers to track network time and uses that information to pinpoint data frame boundaries and slot indexes.[2] It also uses timers to trigger specific tasks as the mobile radio transitions to different operating modes, such as sleep, receive, and talk. A watchdog timer guards against infinite program loops.

The MCU depends on register files to store calibration data, the electronic serial number of the user (used to authenticate the user), and other nonvolatile information.

3.1.2 The DSP

The DSP employs a specialized architecture to handle the vocoder and modulator-demodulator (modem) functions [2]. These functions require tremendous computing power and only recently have become practical for digital systems because of technology and architecture advancements.

The vocoder and modem rely on powerful routines known as algorithms. The algorithms are highly structured and repetitive, making them ideal for software or hardware implementation. The choice of implementation depends on flexibility, speed, and power consumption requirements. In general, hardware algorithms handle chip-rate signal processing, while software algorithms tackle symbol-rate processing. Table 3.1 gives the implementation details of some common algorithms found in CDMA IS95 mobile radios.

The DSP architecture incorporates specialized hardware to efficiently compute certain high-speed functions. The multiply and accumulate structure is one example. It is realized by a parallel multiplier structure or a shift/add structure [3]. Another example is small cache memory for inner-loop instructions [4], and a third example is the correlators used for pilot acquisition and data recovery.

All these hardware improvements reduce the execution time in the DSP. This is crucial because modern communication systems operate with fixed, detailed formats that impose frequent deadlines. As such, it is essential to know the execution time of various signal processing algorithms. That is difficult with a general-purpose processor because it manipulates the flow of data and the instruction sequence to balance loading. In contrast, the DSP uses an explicit instruction set based on very long instruction words (VLIW) [5], which allows hand-crafted code with well-known execution time.

---

2. The slot index identifies the timing associated with slot operation (described in Section 3.2.2).

Table 3.1
MIPS Requirements for Some Common Algorithms Found in a CDMA IS95 Mobile Radio

Algorithm	Implementation	Millions of Instructions Per Second (MIPS)*
Correlator	Hardware	5
Automatic frequency control (AFC)	Hardware	5
Automatic gain control (AGC)	Hardware	5
Transmit filter	Hardware	30
128-pt FFT	Software	1
Viterbi decoder (length = 9, rate = 1/2)	Software	6
Vocoder (8-Kbps Qualcomm code excited linear prediction [QCELP])	Software	20
Vocoder (enhanced variable rate coder [EVRC])	Software	30

*The MIPS values listed describe the performance of a 100-MIPS processor and actually decrease with improved architectures.

3.1.3 Memory

The digital system uses dedicated or shared buses to connect the MCU and the DSP to memory. The memory typically is segmented into blocks that hold the startup code, control software, DSP firmware, and temporary data, as shown in Table 3.2. This approach makes faster access possible, supports zero overhead looping [4], and reduces costs.

3.2 MCU Functions

The MCU serves two main functions, protocol administration and power management.

Table 3.2
Memory Blocks in a Typical Mobile Radio

Memory Block	Function
Boot read-only memory (ROM)	Startup code
Electrical erasable/programmable ROM (EEPROM)	Tuning parameters, user data
Random-access memory (RAM)	DSP firmware, user interface software, and hardware drivers
FLASH (RAM)	Fast access, program, and temporary data

3.2.1 Protocol Administration

The MCU design follows the exact protocol procedures associated with the physical layer, the medium-access control (MAC) layer, and the radio link control layers. The procedures specify network timing, multiple-access approach, modulation format, frame structure, power level, as well as many other details.

The mobile radio attains network synchronization through the pilot and sync channels. The pilot channel is acquired (by aligning the short PN generator of the mobile radio to the received pilot sequence) to establish a link from the base station. That link enables coherent detection and reveals radio propagation effects. The sync channel is decoded to obtain critical timing so that transmitted data packets can be aligned with network frames. That makes it possible to route data through the MAC layer and the radio link control layer.

Call initiation and termination occur through the paging and access channels. The MCU maintains timing during slotted operation, reviews paging channel messages, and directs any network response through the access channel. To make a call or reply to a request, it transmits access probes[3] to draw the attention of the base station and subsequently establish a radio link. To terminate a call, the MCU relays the appropriate signals and powers off key circuits.

The MCU also supervises cell-to-cell handoff through the set maintenance function. This function ensures that the mobile radio connects to the base station with the strongest radio signals. It relies on pilot strength measurements made at the mobile radio to divide the pilot offsets into four categories of decreasing signal strength, as listed in Table 3.3. The active set is especially important because it is the list of pilot signals approved for cell-to-cell handoff.

3.2.2 Power Management

The MCU also provides smart power management to the mobile radio. That includes monitoring battery energy levels, charging the battery, and minimizing power consumption, a vital function for portable equipment.

Table 3.3
Pilot Offset Categories

Category	Description
Active	Recognized and used for handoff
Candidate	Potential active pilot signals
Neighbor	Adjacent base stations and sectors
Remaining	Leftover offsets

---

3. Access probes are signal messages on the access channel.

There are three modes of operation for the mobile phone: idle, receive, and talk. In idle mode, the MCU deactivates most functions except the digital system clock. In receive mode, the MCU activates the RF receiver and the digital modem. It demodulates the paging channel until it receives a valid paging message and then switches to talk mode. In talk mode, the entire mobile radio is active to support two-way communication.

To lower interference, improve system capacity, and extend battery life, networks broadcast paging messages periodically at designated times known as slots, instead of continuously [6]. Slots span two sub-frames (2.5 ms) and occur at multiples of 1.25 sec. Consequently, the mobile radio spends most of the time idle and wakes up the receiver only for the active slots.

Slotted operation complicates mobile radio design, because synchronization is needed to demodulate the received signal; otherwise, the signals are noiselike. Additionally, the received power level is likely to change between slots; therefore, time is needed to adjust the gain of the receiver. As a result, the MCU must queue the receiver before the slot index.

In talk mode, the MCU attains discontinuous transmission by puncturing the RF transmitter. Although this scheme extends battery life, the switching transients actually add more interference to the system.

The mobile radio draws different current levels during slotted and talk modes, as illustrated in Figure 3.2. Furthermore, it is useful to translate the graphs into convenient parameters known as standby time[4] and talk time. Standby time (t_S) is approximated using

$$t_S \approx \frac{\epsilon}{(1 - k)I_{idle} + kI_{Rx}} \tag{3.1}$$

where ϵ is the battery energy in mA·hrs, k is the fraction of time the receiver is on, I_{idle} is the idle current, and I_{Rx} is the receiver current. Talk time (t_T) is estimated using

$$t_T \approx \frac{\epsilon}{I_{Rx} + \nu I_{Tx}} \tag{3.2}$$

where I_{Tx} is the transmitter current and ν is the voice activity factor, typically 3/8. Note that the receiver is always powered on during talk mode.

---

4. Slotted mode operation is more generally referred to as standby time.

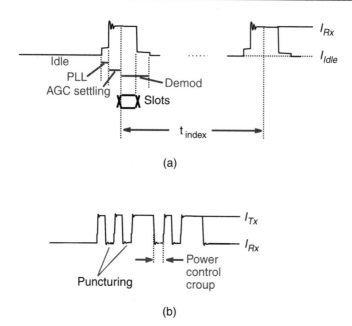

Figure 3.2 Current consumption of the mobile radio during (a) standby time and (b) talk time.

3.3 Digital Signal Processing Algorithms

The DSP executes powerful algorithms on discrete-time digital data. In general, the data is formed by sampling a continuous-time signal and converting the analog samples to digital format. It then becomes possible to perform various digital signal processing routines, including sample rate conversion, digital filtering, spectral estimation (fast Fourier transformation), data windowing, and data detection. The sampling process and the digital signal processing algorithms are outlined below.

3.3.1 The Sampling Theorem

The sampling process converts a continuous analog waveform to discrete-time samples, as illustrated in Figure 3.3. Mathematically, the sampling process modulates the continuous analog signal $x(t)$ by an impulse function $\delta(t - nT)$:

$$y(t) = \sum_{n=-\infty}^{+\infty} x(t)\delta(t - nT) \tag{3.3}$$

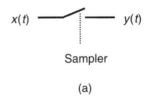

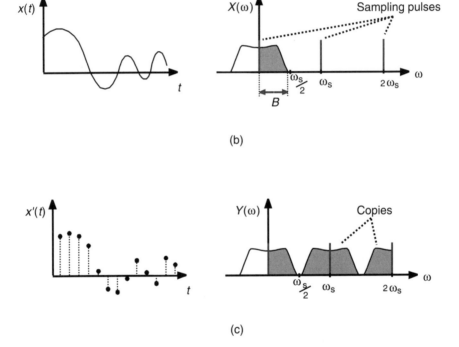

Figure 3.3 Illustration of the sampling process: (a) sampler, (b) input signal, and (c) output signal.

where $y(t)$ are the discrete analog samples and T is the sampling period. The impulse train is a unique function, being periodic in both the time domain and the frequency domain. Its Laplace transform is given by

$$\int_{-\infty}^{\infty} \sum_{n=-\infty}^{\infty} \delta(t - nT)e^{-st}dt = \sum_{n=-\infty}^{\infty} \frac{1}{T}\delta\left[s - \left(\frac{jn}{T}\right)\right] \qquad (3.4)$$

Using that result, the Laplace transform of the sampling process can be written as

$$Y(s) = \int_{-\infty}^{\infty} x(t)e^{-st}dt \; * \; \sum_{n=-\infty}^{+\infty} \frac{1}{T}\delta\left[s - \left(\frac{jn}{T}\right)\right] \tag{3.5}$$

which yields the following important result:

$$Y(s) = \frac{1}{T}\sum_{n=-\infty}^{\infty} X\left[s - \left(\frac{jn}{T}\right)\right] \tag{3.6}$$

The Fourier transform of that result is found by replacing s with $j\omega$:

$$Y(\omega) = \frac{1}{T}\sum_{n=-\infty}^{\infty} X(\omega - n\omega_s) \tag{3.7}$$

where the sampling frequency ω_s is $2\pi/T$. This shows that the frequency spectrum of the sampled signal actually consists of copies of the original signal's spectrum centered at integer multiples of the sampling frequency.

Note that the relationship between the sampling frequency ω_s and the bandwidth of the continuous analog signal $x(t)$ affects the integrity of the sampled signal $Y(s)$. If the sampling frequency is too low, the spectrum copies generated by the sampling process "alias" (overlap) and produce distortion in the sampled signal $y(t)$, as shown in Figure 3.4. By contrast, if the sampling frequency is at least two times the bandwidth of the input signal, the sampled signal retains all the information present in the original signal. This sampling frequency requirement is known as the Nyquist rate [7] and is defined as

$$\omega_s \geq 2B \qquad T \leq \frac{\pi}{B} \tag{3.8}$$

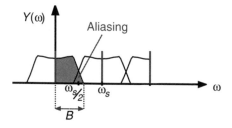

Figure 3.4 An aliasing effect is caused when sampling frequency is too low.

where B is the bandwidth of the original signal. This criterion is fundamental to digital signal processing.

3.3.2 Sample Rate Conversion

The data rate of a signal is an important consideration in digital systems and is chosen for accuracy, convenience, or efficiency. As such, signal processing algorithms often operate at different data rates and thus rely on a resampling process known as sample rate conversion.

The sample rate conversion process is called *decimation* when it lowers the data rate of the original signal and *interpolation* when it raises the data rate of the original signal. The conversion process consists of a sampling operation and linear filtering [8].

The decimation process, shown in Figure 3.5, eliminates samples from the original signal, thereby lowering its data rate and reducing the bandwidth of the decimated signal. This operation "down-samples" the original signal from a rate of T to mT and aliases spurious signals or noise in the input waveform above the new, lower sampling frequency $\omega_s/2m$. To prevent corruption of the decimated signal $y(m)$, the input signal $x(n)$ is passed through a low-pass filter before resampling.

The interpolation process shown in Figure 3.6 adds $k - 1$ zero-valued samples between each pair of the original samples. This operation "up-samples" the input signal $x(n)$ from a rate of T to T/k and creates copies of the original signal spectrum. To remove the copies, the resampled signal is passed through a low-pass filter. The response of the low-pass filter smoothes the zero-valued samples and yields the interpolated signal $y(k)$.

The decimation and interpolation processes include linear low-pass filters. The filters are ideal low-pass filters or box-car filters [9] in the frequency domain and are realized as "accumulate-and-dump" functions, as shown in Figure 3.7.

The accumulate-and-dump function is described by the z-transform [10]:

$$D(z) = 1 + z^{-1} + z^{-2} + \ldots z^{-(N-1)} = \sum_{i=0}^{N-1} z^{-i} \tag{3.9}$$

where z is the unit delay operator equal to e^{-sT}. The z-transform can be rewritten in closed form as

$$D(z) = \frac{1 - z^{-N}}{1 - z^{-1}} \tag{3.10}$$

The frequency response of this filter is found by substituting z for $e^{j\omega T}$:

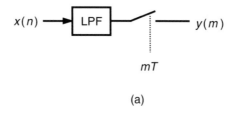

(a)

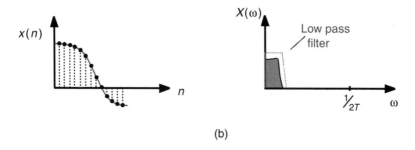

(b)

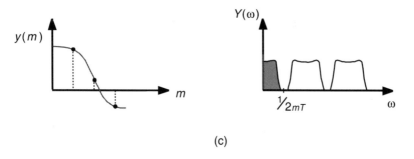

(c)

Figure 3.5 Illustration of the decimation process to reduce sampling rate from *T* to *mT*: (a) block diagram, (b) input signal, and (c) resampled signal.

$$D(\omega) = \left[\frac{1}{N} \frac{\sin\left(\dfrac{\omega N T}{2}\right)}{\sin\left(\dfrac{\omega T}{2}\right)} \right] \tag{3.11}$$

Note that the term $1/N$ vanishes if the filter transfer function is rewritten using sinc functions, where $\mathrm{sinc}(x) = \sin(x)/x$. Accumulate-and-dump filters are also known as comb filters.

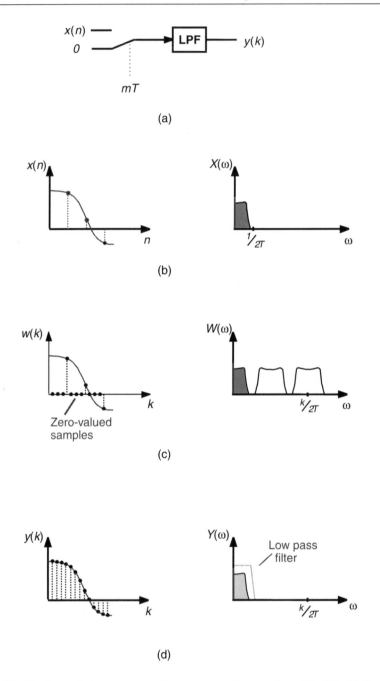

Figure 3.6 The interpolation process to increase sampling rate from T to T/k: (a) block diagram, (b) input signal, (c) effect of zero-value samples, and (d) output signal.

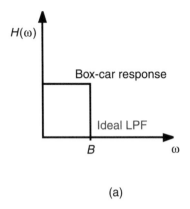

(a)

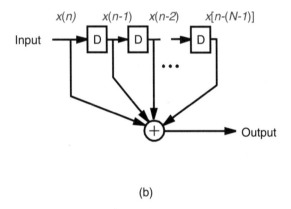

(b)

Figure 3.7 Low-pass filter for sample rate conversion: (a) frequency-domain response of box-car filter and (b) accumulate-and-dump filter structure.

3.3.3 Digital Filters

Digital filters find extensive use in communication systems. They remove interference and noise in the receiver, shape the modulation spectrum prior to transmission, prevent aliasing in sampling operations, enable multirate signal processing, and dampen feedback-control loops. Digital filters generally are linear filters and are classified as finite impulse response (FIR) filters or infinite impulse response (IIR) filters [11].

The FIR filter is a linear constant coefficient filter and is shown in Figure 3.8. It is based on N samples of the input data sequence and is characterized by the transfer function

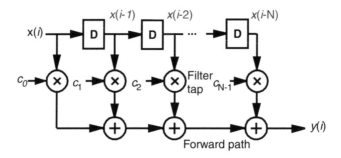

Figure 3.8 Structure of the FIR filter.

$$y(n) = c_0 x(n) + c_1 x(n-1) + \ldots c_{N-1} x(n-N+1) \qquad (3.12)$$

where c_i are the filter tap weights. The filter structure does not include any feedback paths; thus, its transfer function contains only zeros and no poles. Consequently, the FIR filter provides a bounded (unconditionally stable) magnitude response and linear phase response [11].[5] That makes it well suited for phase-modulated communication systems like BPSK and QPSK. The FIR filter is sometimes referred to as a transversal filter.

The IIR filter shown in Figure 3.9 is a recursive digital filter that is similar to traditional analog prototype filters. It includes $N-1$ feedback paths that produce a transfer function with both zeros and poles. As a result, IIR

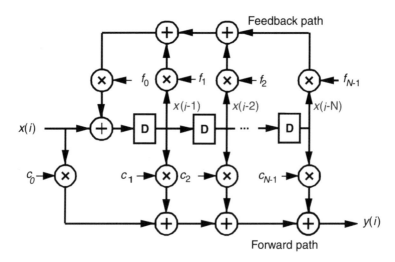

Figure 3.9 Structure of the IIR filter.

5. The linear phase response requires symmetric filter coefficients.

filters are conditionally stable and typically provide high Q responses with steep magnitude transitions and distorted phase responses. IIR filters invariably are more compact and less power hungry than similar performing FIR filters.

3.3.4 Fast Fourier Transforms

Another fairly common function of the DSP is spectral analysis. This analysis is performed by the fast Fourier transform (FFT) algorithm [12], which is simply an efficient procedure to compute the discrete Fourier transform of a data sequence [10].

The discrete Fourier transform (DFT) is defined by

$$X(k) = \sum_{n=0}^{N-1} x(n)e^{\frac{-2\pi jkn}{N}} \tag{3.13}$$

for an N-sample data sequence. It produces N equally spaced components from $-f_s/2$ to $+f_s/2$, where f_s is the sampling frequency. (The frequency components are limited to that range by the Nyquist sampling theorem.) If the shorthand notation W_N is used for the term $e^{-2\pi j/N}$, then

$$X(k) = \sum_{n=0}^{N-1} x(n)W_N^{kn} \tag{3.14}$$

The coefficients of the DFT and the Fourier integral are identical for a band-limited signal sampled at the Nyquist rate. Any difference is due to aliasing distortion caused by too few samples in the data sequence. The DFT also shares many of the useful properties associated with the Fourier transform, including superposition, scaling, time shifting, and convolution [9].

The FFT is a clever computational technique for decomposing and then efficiently rearranging the calculations for the DFT coefficients, thereby speeding computations considerably. This algorithm is realized in one of two basic ways, either decimation in time or decimation in frequency [10].

In both types of FFT algorithms, the computations reduce to a series of two-point DFT operations, known as butterfly computations and shown in Figure 3.10. The butterfly computation is straightforward and computationally simple because it is just a single complex operation. It is described by

$$x_{m+1}(p) = x_m(p) + W_N^r x_m(q) \tag{3.15}$$

$$x_{m+1}(q) = x_m(p) - W_N^r x_m(q)$$

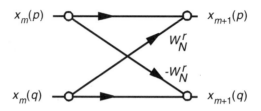

Figure 3.10 The butterfly computation used in the FFT.

As a result, these FFTs require only $(N/2)\log_2 N$ calculations, compared to N^2 calculations for the DFT [10].

3.3.5 Windowing Operations

Signal processing routines typically segment the data stream into data blocks by using windowing operations. The windowing operations either truncate or taper the data sequence; the approach depends on the intended signal processing algorithm. For example, an autocorrelation routine that measures signal power can use a simple windowing function. In contrast, an FFT algorithm requires a periodic data sequence formed by a tapered window.

The windowing operation shown in Figure 3.11 is mathematically described by

$$h(n) = x(n)w(n) \tag{3.16}$$

and can be rewritten as

$$H(\omega) = \sum_n W(\omega)X(\omega - n\omega) \tag{3.17}$$

This second equation, (3.17), is important because it shows that the windowing function alters or smears the frequency spectrum of the data sequence.

The simplest window function is the rectangular window, shown in Figure 3.12 and described by

$$w(n) = 1 \quad \text{for } 0 \leq n \leq N - 1 \tag{3.18}$$
$$= 0 \quad \text{otherwise}$$

The rectangular window is a unit pulse of length N samples, and its discrete Fourier transform is

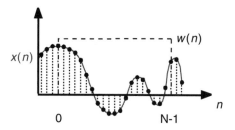

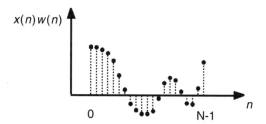

Figure 3.11 Windowing operation.

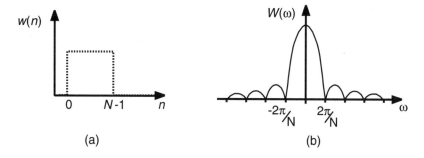

Figure 3.12 Rectangular windowing function in the time domain and frequency domain.

$$W(\omega) = \Im\,[w(n)] = \frac{\sin\left(\dfrac{\omega N}{2}\right)}{\sin\left(\dfrac{\omega}{2}\right)} \tag{3.19}$$

Its spectrum has a main lobe (from $-2\pi/N$ to $2\pi/N$) and several side lobes with measurable energy. The widths of the main lobe and each side lobe depend on the value of N and grow as N becomes smaller. In addition, the energy in each side lobe depends only on the windowing function. As such, several windowing functions have been developed with differing characteristics, as listed in Table 3.4 and shown in Figure 3.13.

3.3.6 Detection Process

The most important function in the communication DSP is the recovery of the transmitted message signal, a process known as detection. The task is a

Table 3.4
Characteristics of Some Popular Windowing Functions [10]

Window	Function		Side Lobe Energy (dB)
Rectangular	$w(n) = 1$ for $0 \le n \le N - 1$		−13
Bartlett	$w(n) = \dfrac{2n}{N-1}$	for $0 \le n \le \dfrac{N-1}{2}$	−25
	$= 2 - \dfrac{2n}{N-1}$	for $\dfrac{N-1}{2} \le n \le N - 1$	
Hanning	$w(n) = \dfrac{1}{2}\left[1 - \cos\left(\dfrac{2\pi n}{N-1}\right)\right]$		−31
Hamming	$w(n) = 0.54 - 0.46\cos\left(\dfrac{2\pi n}{N-1}\right)$		−41
Blackman	$w(n) = 0.42 - 0.5\cos\left(\dfrac{2\pi n}{N-1}\right) + 0.08\cos\left(\dfrac{4\pi n}{N-1}\right)$		−74

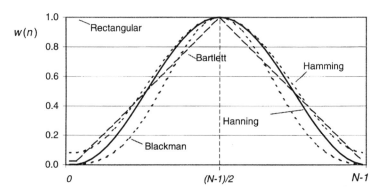

Figure 3.13 Some popular windowing functions.

formidable one for wireless communication systems, because noise and dispersive effects corrupt the link. To combat those effects, the functions of the transmitter and the receiver are designed to complement each other.

Figure 3.14 illustrates a simple matched-filter digital receiver. It consists of a linear filter $h(t)$, a sampler, and a threshold comparator. The transmitted signal $s(t)$ is binary, meaning it has two possible values $s_0(t)$ and $s_1(t)$. The received signal $r(t)$ is given by

$$r(t) = c(t) * s(t) + n(t) \qquad (3.20)$$

where $c(t)$ is the impulse response of the channel and $n(t)$ is white Gaussian-distributed noise. The linear filter $h(t)$ reshapes the received waveform, maximizing the signal energy at the decision points and thereby improving the overall SNR [13].[6]

The sampler produces a single value at $t = T$ described by

$$z(T) = a_i(T) + n(T) \qquad (3.21)$$

where $a_i(T)$ is the signal component of the output, ideally either a_0 or a_1, and $n(T)$ is the noise component of the filter output $z(t)$. The comparator tests $z(T)$ against the threshold γ. If $z(T) < \gamma$, then the hypothesis is h_0, indicating that $s_0(t)$ was sent. Otherwise, $z(T) > \gamma$ and the decision is h_1, suggesting that $s_1(t)$ was transmitted.

The filter $h(t)$ is linear and time invariant. As such, its effect on the Gaussian-distributed input noise $n(t)$ is to produce a second Gaussian random process. Therefore, the noise component $n(T)$ is described by the zero mean process

$$p(n_o) = \frac{1}{\sqrt{2\pi}\sigma} \exp\left(-\frac{n_o^2}{2\sigma^2}\right) \qquad (3.22)$$

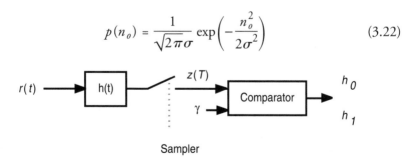

Figure 3.14 Simple digital receiver.

---

6. Assuming that the linear filter $h(t)$ is a matched filter with an inverse transfer function equal to the response of the transmitter plus communication channel.

where n_o is the mean (typically equal to zero) and σ^2 is the noise variance. When added to the signal component $a_i(T)$, a Gaussian random variable, $z(T)$, with a mean of either a_0 or a_1 is produced.

The pdf of $z(T)$ when $s_0(t)$ is transmitted is simply

$$p(z|s_0) = \frac{1}{\sqrt{2\pi}\sigma} \exp\left(-\frac{(z-a_0)^2}{2\sigma^2}\right) \qquad (3.23)$$

which is also known as the conditional probability of $s_0(t)$ given $z(T)$. Similarly, the conditional probability of $s_1(t)$ given $z(T)$ is

$$p(z|s_1) = \frac{1}{\sqrt{2\pi}\sigma} \exp\left(-\frac{(z-a_1)^2}{2\sigma^2}\right) \qquad (3.24)$$

From those conditional pdf's, the optimum comparator is designed, one that minimizes the detection error. In a binary system, a detection error is produced in one of two ways. An error, e, is produced when $s_0(t)$ is transmitted, and the channel noise raises the receiver output signal $z(T)$ above γ. The probability of that is the area under the tail of the Gaussian pdf from γ to ∞, that is,

$$p(e|s_0) = \int_{\gamma}^{\infty} p(z|s_0)dz \qquad (3.25)$$

and is illustrated in Figure 3.15. That expression, also known as the complimentary error function, is defined as

$$erfc[x] = Q[x] = \frac{1}{\sqrt{2\pi}} \int_{x}^{\infty} \exp\left(\frac{-u}{2}\right)^2 du \qquad (3.26)$$

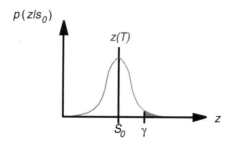

Figure 3.15 Error pdf's for a binary system.

An error, e, is also produced when $s_1(t)$ is transmitted, and the channel noise lowers the receiver output signal $z(T)$ below γ. The probability of that error is

$$p(e|s_1) = \int_{-\infty}^{\gamma} p(z|s_1)dz \qquad (3.27)$$

The probability of bit error p_B is then the sum of those two probabilities, simply

$$p_B = p(e|s_1)p(s_1) + p(e|s_2)p(s_2) \qquad (3.28)$$

Note that both $p(e|s_0)$ and $p(e|s_1)$ are dependent on the Gaussian pdf of the sampled noise $n(T)$ in exactly the same way. Therefore, γ is chosen to make the pdf's symmetric and equal. As a result

$$p_B = p(z|s_0) = p(z|s_1) = Q\left[\frac{a_1 - a_0}{2\sigma}\right] \qquad (3.29)$$

because it is assumed that $p(s_0)$ and $p(s_1)$ are equally likely with value one-half. The best receiver performance is achieved when the threshold γ is set properly, and symbols a_0 and a_1 are spaced as far apart as possible.

An alternative type of digital receiver, which is identical in performance to the matched-filter receiver, is the correlator receiver shown in Figure 3.16. It compares, by cross-correlation, the received signal to each of the possible transmit symbols $s_i(t)$, that is,

$$z_i(t) = \frac{1}{T}\int_0^T r(t)s_i(t)dt \qquad (3.30)$$

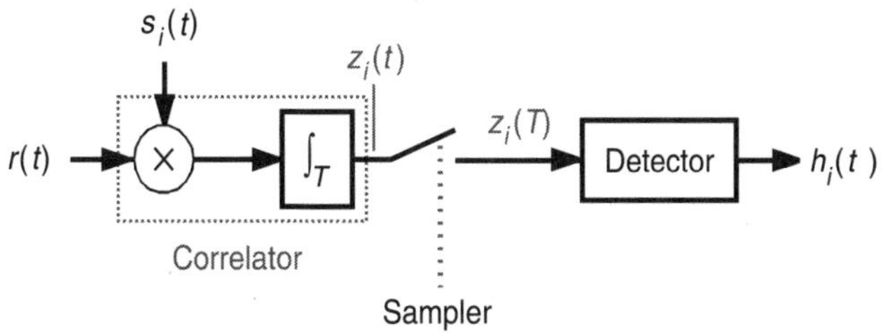

Figure 3.16 Correlator receiver.

where $z_i(T)$ is the decision hypothesis for the ith symbol. The sampler produces a single value at $t = T$ equal to

$$z_i(T) = a_i(T) + n(T) \tag{3.31}$$

which is identical to the matched-filter receiver [13]. This shows that the matched correlator is in fact a synthesis method for the matched-filter receiver. Furthermore, in a binary system, the correlator receiver simplifies to a single correlator function based on either $s_0(t)$ or $s_1(t)$.

For BPSK modulation, the symbols a_0 and a_1 are located at $-\sqrt{E_b}$ and $+\sqrt{E_b}$. It follows that the distance between symbols equals $2\sqrt{E_b}$ and

$$p_B = Q\left(\sqrt{\frac{2E_b}{N_o}}\right) \tag{3.32}$$

where N_o is the noise power. For QPSK modulation, each symbol represents two bits. Because adjacent bits are separated by $\sqrt{2E_s}$ and $E_s = 2E_b$, the distance between adjacent symbols is $2\sqrt{E_b}$. Therefore,

$$p_B = Q\left(\sqrt{\frac{2E_b}{N_o}}\right) \tag{3.33}$$

which is identical to the result for BPSK modulation. This result is striking because it means that the data rate has doubled without affecting the SNR [14].

References

[1] Poor, H. V., and G. W. Wornell (eds.), *Wireless Communications: Signal Processing Perspectives*, Upper Saddle River, NJ: Prentice Hall, 1998.

[2] Stevens, J., "DSPs in Communications," *IEEE Spectrum*, Sept., 1998, pp. 39–46.

[3] Tsividis, Y., and P. Antognetti, (eds.), *Design of MOS VLSI Circuits for Telecommunications*, Englewood Cliffs, NJ: Prentice Hall, 1985.

[4] Eyre, J., and J. Bier, "DSP Processors Hit the Mainstream," *Computer Magazine*, Aug. 1998, pp. 51–59.

[5] Geppert, L., "High-Flying DSP Architectures," *IEEE Spectrum*, Nov. 1998, pp. 53–56.

[6] Garg, V., K. Smolik, and J. E. Wilkes, *Applications of CDMA in Wireless/Personal Communications*, Upper Saddle River, NJ: Prentice Hall, 1997.

[7] Nyquist, H., "Certain Topics in Telegraph Transmission Theory," *AIEE Trans.*, Apr. 1928, pp. 617–644.

[8]   Crochere, R. E., and L. R. Rabiner, "Interpolation and Decimation of Digital Signals—A Tutorial Review," *IEEE Proc.*, Vol. 69, Mar. 1981, pp. 300–331.

[9]   Frerking, M. E., *Digital Signal Processing in Communication Systems*, Norwell, MA: Kluwer Academic Publishers, 1994.

[10]   Oppenheim, A. V., and R. W. Schafer, *Digital Signal Processing*, Englewood Cliffs, NJ: Prentice Hall, 1975.

[11]   Willams, A. B., and F. J. Taylor, *Electronic Filter Design Handbook*, New York: McGraw-Hill, 1995.

[12]   Cochran, W. T., et al., "What Is the Fast Fourier Transform," *IEEE Trans. on Audio and Electroacoustics*, Vol. 15, No. 2, June 1967, pp. 45–55.

[13]   Proakis, J., *Digital Communications*, New York: McGraw-Hill, 1995.

[14]   Couch, L. W., *Digital and Analog Communication Systems*, Upper Saddle River, NJ: Prentice Hall, 1997.

4

Speech Coding

Speech signals are intrinsically analog. Speech signals are converted to digital form to take advantage of the benefits associated with digital communication systems, including access to powerful DSP algorithms, easy interchange of voice and data, message scrambling and encryption, error correction for transmission over noisy channels, and information storage. The conversion process is known as speech coding and is a form of source coding.

Toll-quality[1] digitized speech possesses a high data rate and typically occupies a wider bandwidth than analog speech. The wider bandwidth of digitized speech lowers spectral efficiency, adversely affects system performance, and provides the motivation for data compression and speech coding.

Compression techniques rely on "intelligent" source coding and exploit the perceptive listening qualities of humans. Those qualities give humans the ability to recognize words or phrases spoken by different people, with distinctive voices and accents. Consequently, synthesized speech does not have to duplicate human speech exactly to be easily understood.

This chapter investigates the characteristics of speech, identifies key properties exploited by speech compression algorithms, and presents details of several different coding algorithms. Voice-oriented wireline communication networks typically rely on a technique known as logarithmic pulse code modulation (PCM) for speech coding [1]. In contrast, most wireless communication systems use a form of linear predictive coding (LPC) [1, 2]. This chapter also summarizes the most common methods used to assess the quality of synthesized speech and compares some popular speech-coding algorithms.

---

1. Toll quality relates to the performance found in wireline phone networks.

4.1 Characteristics of Human Speech

Human speech combines two types of sounds, voiced and unvoiced. During voiced sounds, such as vowels, the speaker's vocal chords vibrate at a specific pitch frequency and produce a pulsed output rich in harmonics. That pulsed output or excitation is shaped by the throat, mouth, and nasal passages to form various sounds. During unvoiced sounds, like the consonants *s*, *f*, and *p*, the vocal chords do not vibrate. Instead, turbulent air flow generates a noiselike output that passes through the lips and teeth to create the unvoiced sounds [3, 4].

Human speech is a rich mixture of voiced and unvoiced sound segments, each typically 5–20 ms long and quasistationary [5]. The spectrum of human speech is further characterized by its fine and formant structures, as shown in Figure 4.1. The fine structure is quasiperiodic in frequency and is produced by the vibrating vocal chords. The formant structure is the spectral envelope of the speech signal and is modulated by the vocal tract (i.e., throat, mouth, and nose passages). The spectral envelope shows peaks produced by resonant modes in the vocal tract called formants. A typical speaker demonstrates three

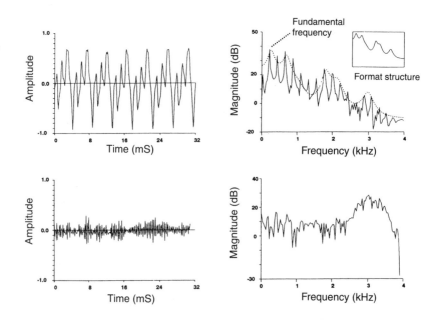

Figure 4.1 Time and frequency domain characteristics of human speech (A. S. Spanias, "Speech Coding: A Tutorial Review," *IEEE Proc.,* © 1994 IEEE).

formants below 3 kHz and one or two additional formants between 3 and 5 kHz [5].

If speech is analyzed over short segments of time, it exhibits several important properties:

- Nonuniform probability density of speech amplitudes;
- Nonflat voice spectra;
- Nonzero autocorrelation function between successive segments [3, 6, 7].

Furthermore, speech can be band-limited[2] without loss of information, making it possible to sample speech at relatively low frequencies and still accurately reproduce it.

In the case of human speech, the nonuniform probability implies that there is a very high probability of near-zero amplitude signals, a significant probability of very high amplitude signals, and a lower probability of in-between amplitude signals. In addition, short-term pdf's are single-peaked and Gaussian-distributed, while long-term pdf's are more likely to be two-sided exponentially flat with a peak at zero, indicating nonspeech. That means the long-term pdf can be approximated by the Laplacian function [3].

The nonflat spectral characteristic of speech follows the individual formant resonances of the speaker. That leads to frequency domain speech-coding algorithms that separate speech into different frequency bands before coding.

The autocorrelation between adjacent speech segments is highly correlated, with typical autocorrelation values of 0.85 to 0.9 [3]. Therefore, a large component of subsequent speech samples can be predicted using the current sample. This concept leads to time-domain predictive algorithms.

4.2 Speech-Coding Algorithms

The speech-coding algorithm encompasses both the encoder, which compresses speech, and the decoder, which synthesizes speech. These functions work together to minimize the transmit bit rate and provide high-quality synthesized speech with low complexity and low delay.

Speech-coding algorithms generally fall into one of two categories, waveform coders or vocoders. Waveform coders focus on the speech waveform,

---

2. Most wireline and wireless digital communication systems band-limit speech to approximately 3.6 kHz and sample it at an 8-kHz rate.

using scalar and vector quantization methods to faithfully reconstruct the speech signal. Waveform coders are generally robust and are suitable for a wide class of signals. Vocoders take advantage of speech characteristics to produce perceptually intelligible sounds without necessarily matching the original speech waveform. As such, they are suited for speech-only low-rate applications, like digital cellular telephones.

All speech-coding algorithms begin by converting analog speech to digital form. The conversion process periodically samples the continuous analog signal and maps the samples to a set of discrete codes. As a result, the quantization process introduces irreversible distortion because each unique code word represents a range of analog values.

In a standard A/D converter (or quantizer), the discrete codes are uniformly spaced. Furthermore, to achieve high-quality digitized speech, the analog input is typically sampled at an 8-kHz rate with 13-bit resolution.

4.2.1 Waveform Coders

Waveform coders support general-purpose applications, provide medium-rate (16 to 64 Kbps) performance with above-average quality, and find widespread use in wireline communication systems. Waveform coders use quantizers to produce an output data stream with binary values that appears pulselike (hence, the label *pulse code modulation*). A standard quantizer generates uniform PCM, runs at full rate (104 Kbps), and possesses a wide bandwidth.

To reduce the output bit rate (and bandwidth) of the PCM coder, a nonuniform quantizer is used. In this type of coder, the quantization levels are fine for frequently occurring signal amplitudes and coarse for rarely occurring signal amplitudes. The quantizer levels are typically spaced using one of two near-logarithmic functions. One of these is μ-law companding [6] and is based on the expression

$$V_{out}(t) = \frac{\ln(1 + \mu |V_n(t)|)}{\ln(1 + \mu)} \tag{4.1}$$

where $V_n(t)$ is normalized using $V_{in}(t)/V_{max}$ and $\mu = 255$. The other is A-law companding [6] and is based on these equations:

$$V_{out}(t) = \frac{A(|V_n(t)|)}{1 + \ln(A)} \qquad \text{for } 0 < |V_n(t)| < \frac{1}{A} \tag{4.2a}$$

$$V_{out}(t) = \frac{1 + \ln[A|V_n(t)|]}{1 + \ln(A)} \qquad \text{for } \frac{1}{A} < |V_n(t)| < 1 \tag{4.2b}$$

where A = 87.56. These functions, which are graphed in Figure 4.2, are effective at compressing 13-bit signals (104-Kbps data rate) to 8-bit format (64 Kbps), the data rate used by most wireline communication networks.

There are other scalar quantization schemes that further reduce the data rate. In adaptive PCM (APCM) [5], the dynamic range of the quantizer tracks the amplitude of the signal. It uses the time-varying property of speech signals and relies on the amplitude of the previous sampled signal to set the range of the quantizer. A more efficient method, known as differential PCM (DPCM) [8] and shown in Figure 4.3, further exploits the correlation between adjacent samples.

Practical DPCM coders include a time-invariant short-term predictor to estimate the current speech sample, $s(n)$. It forms an estimate, $\hat{s}(n)$, using p past samples and the following relationship:

$$\hat{s}(n) = a_1 s(n-1) + a_2 s(n-2) + \ldots = \sum_{k=1}^{p} a_k s(n-k) \qquad (4.3)$$

with fixed coefficients $\{a_k\}$. That shows that the predictor is simply a linear FIR filter with p taps. The estimate can also be written in z-notation as

$$\hat{S}(z) = \sum_{k=1}^{p} a_k S(z) z^{-k} = [1 - A(z)]S(z) \qquad (4.4)$$

where z denotes the z-transform.

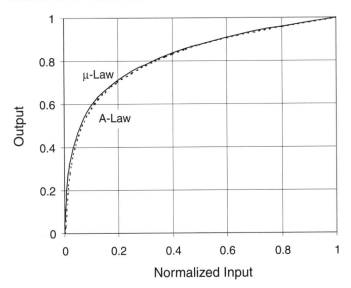

Figure 4.2 Log-PCM companding.

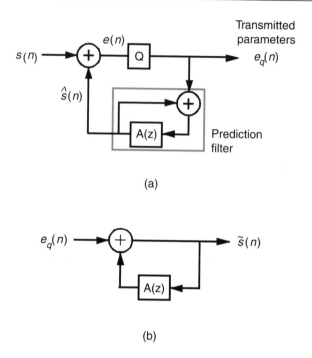

Figure 4.3 DPCM: (a) encoder and (b) decoder.

The DPCM coder determines the difference, $e(n)$, between the current speech sample and the estimate, $s(n) - \hat{s}(n)$. It codes that information and then transmits it to the receiver. At the receiver, the decoder applies the information to a matching prediction filter to synthesize the speech sample.

A subclass of DPCM is delta modulation [8], which operates at a much higher rate but uses a single bit to represent the prediction error. Another subclass of DPCM is adaptive DPCM (ADPCM) [9, 10]. It allows the step size and the predictor coefficients to vary and track the speech input.

4.2.2 Vocoders

For wireless networks, it is advantageous to further compress the data stream and thereby make more efficient use of the radio spectrum. That is accomplished with vocoder algorithms, which exploit the characteristics of human speech. The algorithms compress the data rate to 4 to 16 Kbps with acceptable complexity and toll quality.

Figure 4.4 illustrates the vocoder concept, which is modeled after human speech physiology. It simulates voiced sounds by a periodic impulse generator at the fine structure frequency and unvoiced sounds by a noise source. The

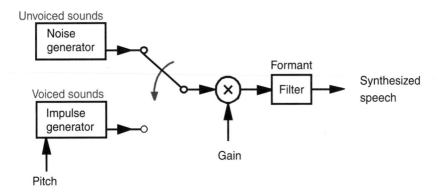

Figure 4.4 Vocoder model for generation of synthetic speech.

signals pass through a gain element to adjust the energy level of the signal and a formant filter to represent the effect of the vocal tract.

The vocoder algorithm divides speech segments into long (unvoiced) and short (voiced) events. It then processes those signals in such a way as to map more bits to rapidly changing elements and fewer bits to slowly changing elements. There are several different vocoder algorithms, which are outlined next.

4.2.2.1 Channel Coders

The channel coder is a simple frequency domain vocoder that exploits the nonflat spectral characteristics of speech. It measures the spectral envelope (formant structure) of the speech signal by separating the signal into frequency bands using the structure shown in Figure 4.5. The structure typically consists of 16 to 19 channel filters with increasing bandwidth at higher channel frequency [5]. The channel coder samples the energy in each frequency band every 10 to 30 ms.

The channel coder also analyzes the fine structure of the speech sequence and determines the characteristics of the excitation source, including the gain factor, the binary voice decision, and the fundamental pitch frequency. The gain factor scales the resulting coded speech to match the total energy level of the input speech. The binary voice decision specifies the appropriate excitation source. If the sound segment is "noiselike" with low energy and a large number of zero crossings, then it is "unvoiced" sound and simulated by a random noise generator. Otherwise, the sound segment is "voiced," with a fundamental pitch frequency (found by recognizing peaks in the autocorrelation sequence) and simulated by a periodic pulse generator.

The channel coder transmits compressed binary data that describe the voice excitation and spectral envelope (as measured by the formant structure)

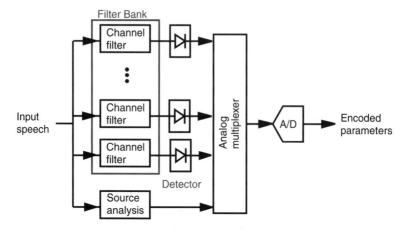

Figure 4.5 Block diagram of a channel coder.

of the input speech sequence. The decoder uses those parameters to reconstruct the speech sequence in the frequency domain.

4.2.2.2 Linear Predictive Coders

The LPC [3, 4, 11–13] is a time domain vocoder, which improves on the performance of the channel coder by replacing the channel filters with a more versatile filter, as shown in Figure 4.6. It extracts significant features of the speech signal (such as the spectral envelope, the pitch frequency, and the energy level), codes those parameters, and transmits them to the receiver, where the speech signal is synthesized. The LPC is computationally intensive but has become practical with the development of DSP architectures and CMOS VLSI technology.

Speech synthesis is modeled after Figure 4.4 and described by

$$S(z) = V(z)\frac{G}{A(z)} \qquad . \qquad (4.5)$$

where $S(z)$ is the synthesized speech signal, $V(z)$ is the excitation, G is the gain factor, and $1/A(z)$ is called the synthesis filter. Equation (4.5) is rearranged to read

$$S(z) = [1 - A(z)]S(z) + GV(z) \qquad (4.6)$$

and then transformed to the time domain, where

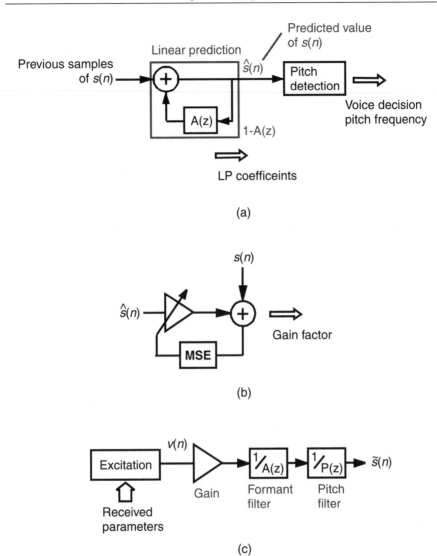

Figure 4.6 LPC: (a) spectral envelope and pitch determination, (b) gain factor analysis, and (c) decoder.

$$s(n) = \sum_{k=1}^{p} a_k s(n-k) + G v(n) \qquad (4.7)$$

Note that the first term forms an all-pole linear filter and shapes the spectral envelope of the speech signal.

The coefficients for the synthesis filter are estimated using linear prediction. In that approach, the current sample estimate is the linear sum of p (typically 8 to 16 [3]) past samples and is written as

$$\hat{s}(n) = \sum_{k=1}^{p} \alpha_k s(n - k) \tag{4.8}$$

where $\{\alpha_k\}$ are the adaptive filter coefficients or estimates. Unfortunately, this filter does not perfectly match the vocal tract and therefore produces a prediction residual, $e(n)$, given by

$$e(n) = s(n) - \hat{s}(n) = s(n) - \sum_{k=1}^{p} \alpha_k s(n - k) \tag{4.9}$$

It is important to note that if $a_k = \alpha_k$, then the prediction residual is simply the excitation source $Gv(n)$.

The adaptive filter coefficients are determined by minimizing the prediction residual. That method uses the average energy E in the error signal, which is given by

$$E = \sum_{n=1}^{N} e^2(n) = \sum_{n=1}^{N} \left[s(n) - \sum_{k=1}^{p} \alpha_k s(n - k) \right]^2 \tag{4.10}$$

where N is chosen as a compromise between accuracy and expected autocorrelation properties [3].[3] To find the filter coefficients, the expression for the average energy in the error signal is differentiated with respect to the filter coefficients α_m and set to zero. That yields

$$\frac{\partial E}{\partial \alpha_m} = \sum_{k=0}^{p} C_{mk} \alpha_k = 0 \tag{4.11}$$

where C_{mk} describes the correlation between the sample $s(n - m)$ and the other weighted samples $\alpha_k s(n - k)$ [4, 5, 11]. The linear equations for the predictor coefficients are called normal equations or Yule-Walker equations and are solved efficiently by the Levinson-Durbin algorithm [12]. Ideally, the prediction filter acts as a short-term decorrelator and produces an error residual that has a flat power spectrum.

---

3. The autocorrelation of the signal as a function of delay indicates how quickly the LPC changes.

Recall that the synthesis filter was defined as $1/A(z)$. Consequently, an error in any one coefficient affects the entire frequency spectrum, which can have disastrous consequences on the quality of the synthesized speech. That shortcoming can be removed by transforming the adaptive coefficients $\{\alpha_k\}$ to zeros in the z-plane. With this approach, each pair of zeros describes a resonant frequency with a resonant bandwidth. The zeros of the z-transform are called the line spectrum frequencies and are grouped to form line spectrum pairs (LSPs). In practice, the LSP parameters are more immune to errors[4] and are normally quantized and transmitted instead of the adaptive filter coefficients [13, 14].

The classical linear predictive coder described so far uses a synthesis-and-analysis approach without feedback. It transmits features of the prediction residual or error signal to excite the synthesis filter at the receiver. (That contrasts with the DPCM coder, which actually transmits the quantized error signal.) Its performance is affected by the accuracy of the prediction filter and the excitation source. Reliable estimation of the spectral envelope is possible using linear prediction techniques, but accurate estimation of the excitation source is more challenging. At low data rates, the basic two-state excitation source, consisting of a white noise generator (for unvoiced sounds) and a variable-rate pulse generator (for voiced sounds), produces synthetic speech quality. To achieve better performance, closed-loop linear predictive coders that optimize more flexible excitation sources are used.

In closed-loop form, the linear predictive coder drives the difference between the synthesized waveform $\tilde{s}(n)$ and the input signal $s(n)$ toward zero. This type of vocoder uses synthesis by analysis to optimize the excitation source, as shown in Figure 4.7. It produces an error signal given by

$$e(n) = s(n) - \tilde{s}(n) \tag{4.12}$$

and includes an algorithm to select the best excitation source. The choice of error minimization criterion is critical and is typically based on the mean square error (MSE) between the input and the synthesized sequences:

$$E = \sum_n e^2(n) = \sum_n \left[s(n) - \sum_{k=1}^{p} \alpha_k s(n-k) \right]^2 \tag{4.13}$$

Other error-minimization methods are based on autocorrelation and autocovariance functions [3, 4, 11].

---

4. It is easier to guarantee the stability of the synthesis filter using LPC parameters.

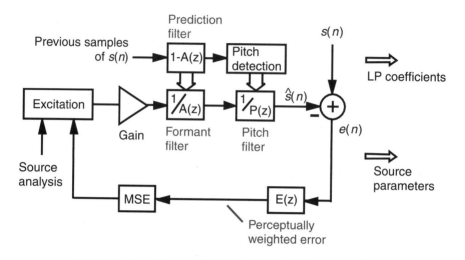

Figure 4.7 Analysis-by-synthesis linear predictive coder.

A straightforward error criterion, like MSE, does not account for the perceptual quality of digitized speech. That perspective is gained by adding a weighted filter $E(z)$ to shape the error spectrum and concentrate energy at the formant frequencies. As a result, the errors at the formant frequencies are minimized. In practice, the weighted filter is a model of the human ear's response.

4.2.2.3 RPE-LTP Algorithm

The GSM system uses the regular pulse excitation (RPE)-long-term predictor (LTP) speech-coding algorithm [15] shown in Figure 4.8. The algorithm consists of a linear prediction filter and adds an advanced multipulse excitation

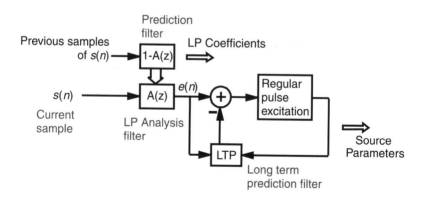

Figure 4.8 Block diagram of the RPE-LTP algorithm used in GSM systems.

source. It models the voice source with the RPE technique and analyzes the pitch frequency with an LTP. The LTP uses an adaptive, single-tap filter to estimate the pitch of the linear prediction residual, $s(n) - \tilde{s}(n)$. The resulting pitch frequency identifies three candidate excitation sequences, which are analyzed to find the best match.

The RPE-LTP performs nearly as well open-loop as it does closed-loop. That is, the residual analysis can directly select the excitation sequence without significant loss of information. That simplification reduces the complexity of the algorithm, although it still remains relatively high. The RPE-LTP algorithm compresses speech to a data rate of 13 Kbps with average quality.

4.2.2.4 Code Excited LPC Algorithms [16]

In some speech-coding algorithms, the excitation source is realized by a codebook, or list of excitation waveforms, as shown in Figure 4.9. The codebook is used to store stochastic, zero-mean, white, Gaussian excitation signals that are common to both the encoder and the decoder. For each speech segment, the codebook is searched for the best perceptual match and the corresponding index is transmitted. The coders are extremely complex but are superior to the quality of two-state versions.

The vector summed excitation linear predictive (VSELP) algorithm is used by North American digital cellular (NADC) systems [17, 18]. The excitation is generated from the vector sum of three basis vectors, consisting of an adaptive codebook to realize the long-term prediction (pitch) filter and two VSELP codebooks. Each basis vector is orthogonal to the other two, which facilitates joint optimization. The complexity is further reduced because the vector sum

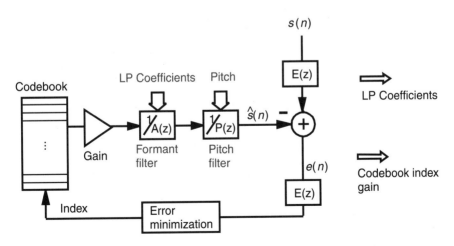

Figure 4.9 Block diagram of the codebook excited linear prediction (CELP) coder.

is constrained to simple addition or subtraction of the basis vectors and the number of basis vectors is relatively few. The VSELP algorithm provides compressed data at a rate of 8 Kbps with slightly better quality than the RPE-LTP algorithm.

Another code excited linear predictive coding algorithm is the QCELP algorithm developed for CDMA IS95 [19]. It integrates four coding rates with a scalable architecture to achieve variable-rate speech coding. The four rates enable lower average data rates, which translate to lower average transmit power levels, thereby minimizing network interference. The algorithm sets the coding rate (full, half, quarter, eighth) by comparing the energy of each 20-ms speech segment against three energy levels.

The three energy levels adjust dynamically to account for changes in background noise and speaker volume. That allows efficient coding of the linear prediction coefficients, gain factor, and excitation source.

The QCELP codebook is circular and based on a 128-by-128 matrix. The circular design means the next entry is merely the current entry shifted by one sample. That allows the entire codebook to be stored as a single 128-sample vector. As a result, these benefits are realized: smaller memory size plus accelerated and simplified digital signal processing.

The QCELP speech-coding process is as follows. First the LP filter coefficients $A(z)$ are determined and translated into LSPs. The LP filter is also known as the formant filter and is used to remove the short-term correlation in the digitized speech. The next step is to estimate the long-term predictor coefficients $H(z)$ that correspond to the pitch frequency of the digitized speech and determine the coding rate for the frame. The codebook excitation is then found by minimizing the weighted error between the input speech and the coded speech, which results when the excitation source passes through the pitch and formant filters.

The LP filter coefficients and voice activity rate are found using the autocorrelation function, given as

$$R(k) = \sum_{n=1}^{N-k} s(n)s(n+k) \tag{4.14}$$

where $s(n)$ represents the windowed version of the input sequence and N equals 160, corresponding to the number of samples in a frame. The first 10 autocorrelation results[5] are input to the Levinson-Durbin linear prediction algorithm to determine the appropriate filter coefficients. The first 16 coefficients are analyzed for rate determination.

---

5. The results are normally referred to as log-area ratios (LARs) [4, 5, 11].

Rate determination is a two-step process. First, the binary voice decision (voiced or unvoiced speech) is made. Voiced sounds are mapped to either full- or half-rate, while unvoiced sounds are assigned to eighth-rate. In the second step, the voiced sound is reviewed more closely and categorized as either full-, half-, or quarter-rate. Full-rate coding is used for transitional speech, that is, sounds that change during the frame. Dynamic thresholds are used to account for background noise changes, as shown in Figure 4.10.

Variable-rate operation is achieved by reducing the size and the length of the codebook and the linear prediction quantizer. At full- and half-rates, the full codebook is used, while at quarter- and eighth-rates, a pseudorandom vector generator is substituted.

The weighting filter used by the QCELP algorithm is relatively simple: It is related to the formant filter and is described by

$$E(z) = \frac{A(z)}{A\left(\frac{z}{\zeta}\right)} \qquad (4.15)$$

where $\zeta = 0.78$ [1, 4, 21]. This concept exploits the physiology of the vocal tract to perceptually shape the error.

The QCELP encoder assigns bits to each 20-ms frame, as shown in Figure 4.11. For slow-varying speech, the encoder requires fewer parameters to describe the excitation source and the filters. For fast-varying speech, the encoder analyzes digitized speech in subframes as small as 2.5 ms. In practice, full- and eighth-rates occur more often than the other rates.

The CELP algorithm is improved with the enhanced variable rate coder (EVRC) [22]. It combines the advantages of two algorithms. It uses the relaxation CELP algorithm [23] to find the prediction residual and the algebraic

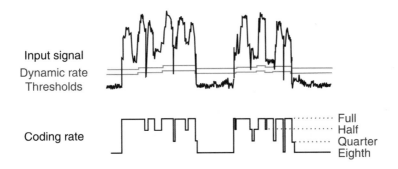

Figure 4.10 The speech-coding rate depends on the dynamic variations of the speech waveform [20].

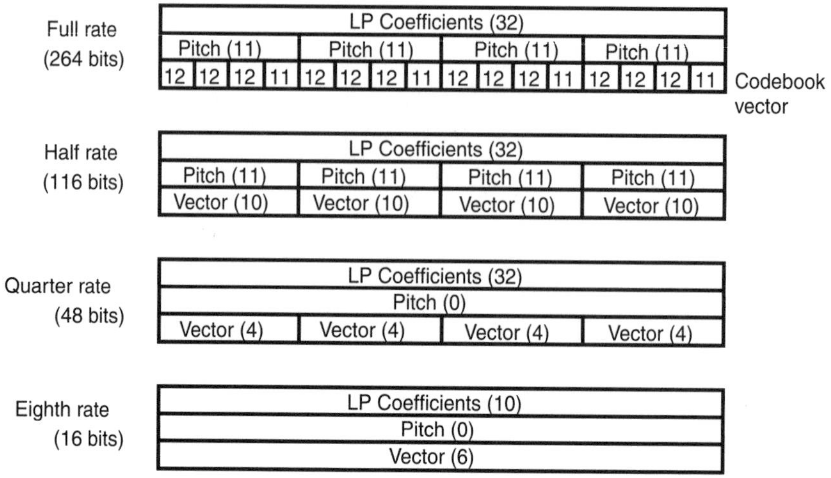

Full rate
(264 bits)

LP Coefficients (32)			
Pitch (11)	Pitch (11)	Pitch (11)	Pitch (11)
12	12	12	11	12	12	12	11	12	12	12	11	12	12	12	11

Codebook vector

Half rate
(116 bits)

LP Coefficients (32)			
Pitch (11)	Pitch (11)	Pitch (11)	Pitch (11)
Vector (10)	Vector (10)	Vector (10)	Vector (10)

Quarter rate
(48 bits)

LP Coefficients (32)			
Pitch (0)			
Vector (4)	Vector (4)	Vector (4)	Vector (4)

Eighth rate
(16 bits)

LP Coefficients (10)
Pitch (0)
Vector (6)

Figure 4.11 Frame structure of digitized speech in CDMA IS95 systems.

CELP algorithm to code the residual with low complexity.[6] A relaxed error minimization criterion in the feedback loop (that shapes the match between the original speech and synthesized speech) is key to this speech-coding algorithm. It allows the pitch period to be coded once per frame, while linear interpolation techniques estimate the pitch period of each subframe. That can lead to a large mismatch between the input and coded speech sequences. To correct the mismatch, the original residual is modified to match a time-warped version of the speech signal. The key result is a reduced coding rate without significant loss in perceptual quality.

In the EVRC, every frame is split into three subframes, and the codebook is searched during each of those subframes. The search targets full-, half-, and eighth-data rates. For full-rate speech, the excitation consists of eight pulses defined in position by a 35-bit codebook. For half-rate speech, the excitation uses three pulses and a 10-bit codebook. At eighth-rate, the excitation source is identical to that of the QCELP coder and is a pseudorandom vector generator.

4.2.3 Speech Coders for Wireless Communication Systems

Table 4.1 summarizes the LPC algorithms commonly used by wireless communication systems.

---

6. Low complexity translates to small memory size for the codebook and low MIPS requirements for the DSP.

Table 4.1
Compression Algorithms for Leading Wireless Standards [2, 3]

Standard	Coding Algorithm	Compressed Rate (Kbps)
GSM	RPE-LTP vocoder	13
NADC	VSELP vocoder	8
PHS	ADPCM	32
CDMA IS95	QCELP vocoder	0.8–13.3
	EVRC	0.8–8.5

4.3 Speech Quality

Speech quality is degraded during quantization and further compromised by data compression. Those effects are unavoidable, but they are outweighed by the well-known benefits of digital communications and are essential in all but friendly wireless environments.

Speech-coding algorithms are evaluated based on the bit rate, algorithm complexity, delay, and reconstructed speech quality. Minimizing the bit rate is a primary concern in a wireless communication system, because it is directly linked to the bandwidth of the radio signal. Minimizing the algorithm complexity is vital, because it burdens the DSP and drains battery energy. Some vocoder algorithms, such as CELP, operate at 20 MIPS. The delay of the system, including source and channel coding, is limited to less than about 50 ms: otherwise, the delay is noticeable to the user [3]. The speech quality is targeted to be at wireline, or toll, quality.

Evaluating the quality of digitized speech is a difficult task, in part because subtle features of the speech waveform have a significant impact on its perceptual quality. Digitized speech typically is classified into one of four categories of decreasing quality: broadcast, network or toll, communications, and synthetic [5].

Speech quality is analyzed in a variety of ways. One common technique is to measure the SNR of the synthesized speech sequence, which is described by

$$\text{SNR} = 10\log_{10}\left\{ \frac{\sum_{n=0}^{M-1} s^2(n)}{\sum_{n=0}^{M-1} [s(n) - \hat{s}(n)]^2} \right\} \tag{4.16}$$

where $s(n)$ is the original speech waveform and $\tilde{s}(n)$ is the coded speech data. This is a long-term measurement that hides temporal variations in the speech waveform. A short-term measure that sums the SNR performance of smaller speech periods is the segmented SNR (SEGSNR) [5], which is given by

$$\text{SEGSNR} = \frac{10}{L} \sum_{i=0}^{L-1} \log_{10} \left\{ \frac{\displaystyle\sum_{n=0}^{N-1} s^2(iN + n)}{\displaystyle\sum_{n=0}^{N-1} [s(iN + n) - \hat{s}(iN + n)]^2} \right\} \qquad (4.17)$$

where N typically covers 5 ms. It exposes weak signals and generally provides a better performance measure. Other objective measures include the articulation index, the log spectral distance, and the Euclidean distance.

Most speech coders are based on perceptual encoding and as such are better judged by subjective methods. There are three commonly used subjective methods: the diagnostic rhyme test (DRT), the diagnostic acceptability measure (DAM), and the mean opinion score (MOS) [24]. For the diagnostic acceptability measure [25], the listener is asked to recognize one of two words in a set of rhyming pairs. For the diagnostic rhyme test [26], a trained listener scores speech quality using a normalized reference. With the mean opinion score [27], the naive listener's opinion of the reconstructed speech, using the scale shown in Table 4.2, is recorded. The last method is the most popular because it does not require any reference, although it is highly subjective. In the DRT and mean opinion score methods, several listeners are screened.

Table 4.3 lists the subjective qualities for leading speech-coding algorithms. As expected, the speech quality degrades as the bit rate decreases.

Table 4.2
Mean Opinion Score Quality Rating [27]

Quality	Scale	Listening Effort
Excellent	5	No effort
Good	4	No appreciable effort
Fair	3	Moderate effort
Poor	2	Considerable effort
Bad	1	Not understood

Table 4.3
Mean Opinion Scores for Some Popular Coders [28, 29]

Coder	Mean Opinion Score
64-Kbps PCM	4.3
13-Kbps QCELP	4.2
32-Kbps ADPCM	4.1
8-Kbps VSELP	3.7
13-Kbps RPE-LTP	3.54

References

[1]   Budagavi, M., and J. D. Gibson, "Speech Coding in Mobile Radio Communications," *IEEE Proc.*, July 1998, pp. 1402–1412.

[2]   Steele, R., "Speech Codecs for Personal Communications," *IEEE Communications Magazine*, Nov. 1993, pp. 76–83.

[3]   Rappaport, T. S., *Wireless Communications: Principles and Practice*, Upper Saddle River, NJ: Prentice Hall, 1996.

[4]   Frerking, M. E., *Digital Signal Processing in Communication Systems*, Norwell, MA: Kluwer Academic Publishers, 1994.

[5]   Spanias, A. S., "Speech Coding: A Tutorial Review," *IEEE Proc.*, Oct. 1994, pp. 1541–1582.

[6]   Flanagan, J. L., et al., "Speech Coding," *IEEE Trans. on Communications*, Vol. COM-27, No. 4, Apr. 1979, pp. 710–735.

[7]   Atal, B. S., and M. R. Schroeder, "Stochastic Coding of Speech Signals at Very Low Bit Rates," *Proc. IEEE International Conf. on Communications*, 1984, pp. 1610–1613.

[8]   Jayant, N. S., "Digital Coding of Speech Waveforms: PCM, DPCM, and DM Quantizers," *IEEE Proc.*, Vol. 62, May 1974, pp. 611–632.

[9]   Cummiskey, P., et al., "Adaptive Quantization in Differential PCM Coding of Speech," *Bell Systems Tech. J.*, Vol. 52, No. 7, Sept. 1973.

[10] Gibson, J., "Adaptive Prediction in Speech Differential Encoding Systems," *IEEE Proc.*, Vol. 68, Nov. 1974, pp. 1789–1797.

[11] Steele, R. (ed.), *Mobile Radio Communications*, Chichester, Eng.: Wiley, 1996.

[12] Proakis, J. G., *Digital Communications*, New York: McGraw-Hill, 1995.

[13] Itakura, F., and S. Saito, "On the Optimum Quantization of Feature Parameters in the PARCOR Speech Synthesizer," *IEEE Conf. on Speech Communications and Processing*, Apr. 1972, pp. 434–437.

[14] Viswanatham, V., and J. Makhoul, "Quantization Properties of Transmission Parameters in Linear Predictive Systems," *IEEE Trans. on Acoustics, Speech, and Signal Processing*, Vol. ASSP-23, June 1975, pp. 309–321.

[15] Kroon, P., E. F. Deprettere, and R. J. Sluyter, "Regular Pulse Excitation—A Novel Approach to Effective and Efficient Multi-Pulse Coding of Speech," *IEEE Trans. on Acoustics, Speech, and Signal Processing*, Vol. ASSP-24, Oct. 1986, pp. 1054–1063.

[16] Schroeder, M. R., and B. S. Atal, "Code-Excited Linear Prediction (CELP): High-Quality Speech at Very Low Bit Rates," *IEEE*, 1985.

[17] Mermelstein, P., "The IS-54 Digital Cellular Standard," in J. D. Gibson (ed.), *The Communications Handbook*, Boca Raton, FL: CRC Press, 1997, pp. 1247–1256.

[18] Gerson, I., and M. Jasiuk, "Vector Sum Excited Linear Prediction (VSELP) Speech Coding at 8 Kbits/s," *Proc. ICASSP-90*, Apr. 1990, pp. 461–464.

[19] EIA/TIA, "Speech Service Option Standard for Wideband Spread Spectrum Digital Cellular Systems," IS-96A, May, 1995.

[20] Leonard, M., "Digital Domain Invades Cellular Communications," *Electronic Design*, Sept. 17, 1992, pp. 40–52.

[21] Wang, D. Q., "QCELP Vocoders in CDMA Systems Design," *Communications Systems Design Magazine*, Apr. 1999, pp. 40–45.

[22] EIA/TIA, "Enhanced Variable Rate Codec, Speech Service Option 3 for Wideband Spread Spectrum Digital Systems," IS-127, Sept. 9, 1996.

[23] Kleijn, W. B., and W. Granzow, "Methods for Waveform Interpolation in Speech Coding," *Digital Signal Processing*, Vol. 1, No. 4, 1991, pp. 215–230.

[24] Kubichek, R., "Standards and Technology Issues in Objective Voice Quality Assessment," *Digital Signal Processing: Rev. J.*, Vol. DSP 1, Apr. 1991, pp. 38–44.

[25] Fairbanks, G., "Test of Phonemic Differentiation: The Rhyme Test," *J. Acoustic Society of America*, Vol. 30, 1958, pp. 596–600.

[26] Voiers, W. D., "Diagnostic Acceptability Measure for Speech Communication Systems," *Proc. ICASSP*, May 1977.

[27] Quackenbush, S. R., T. P. Barnwell, and M. A. Clements, *Objective Measures for Speech Quality*, Englewood Cliffs, NJ: Prentice Hall, 1988.

[28] Coleman, A., et al., "Subjective Performance Evaluation of the RPE-LTP Codec for the Pan-European Cellular Digital Mobile Radio System," *Proc. ICASSP*, 1989, pp. 1075–1079.

[29] Jayant, N. S., "High Quality Coding of Telephone Speech and Wideband Audio," *IEEE Communications Magazine*, Jan. 1990, pp. 10–19.

5

Digital Modem

Robust communication over a wireless channel requires conditioning of the message signal against fading and interference. That often means increasing the signal's bandwidth to achieve some improvement in system performance. For direct-sequence spread-spectrum communication systems, the conditioning significantly increases the signal's bandwidth, but it also enables multiple users to share the radio channel simultaneously.

Modern wireless communication systems condition the message signal at the transmitter and recover the signal at the receiver using powerful DSP algorithms. The algorithms are executed by the digital modulator and demodulator, known together as the digital modem and shown in Figure 5.1. The modulator superimposes the message waveform onto a carrier for radio transmission. It uses methods that guard against fading and other impairments while it maximizes bandwidth efficiency. The demodulator detects and recovers the transmitted digital message. It tracks the received signal, rejects interference, and extracts the message data from noisy signals.

This chapter investigates the modem in the mobile radio, covering some general design issues and concentrating on specific CDMA IS95 implementations. It describes the key operations in the digital modulator: synchronization, channel coding, and signal filtering. The chapter also presents the algorithms performed by the digital demodulator, including pilot acquisition, carrier recovery, AGC, data detection, and data recovery.

5.1 Digital Modulator

The digital modulator in the mobile radio codes the message data for transmission over the reverse-link wireless channel and detection at the base station.

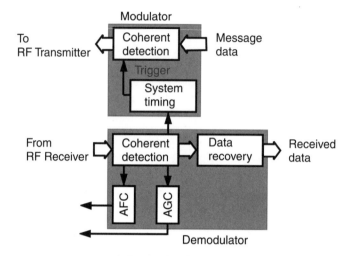

Figure 5.1 The digital modem.

The reverse link in CDMA IS95 suffers on two accounts. First, the mobile radio does not transmit the chip-rate timing signal and thereby requires the base station receiver to reconstruct the phase reference. Such a detection method lowers performance by as much as 3 dB when compared to coherent detection methods in AWGN channels [1]. Second, the base station receives signals from mobile radios that are randomly placed and often moving. That makes it difficult to synchronize the received signals, which is important for orthogonal spreading, and to accurately control the received power levels, which is essential for realizing maximum user capacity.

The digital modulator shown in Figure 5.2 conditions the signal to improve the detection process through coarse-timing synchronization[1] and fast power control. It aligns, formats, and modulates data for transmission using methods outlined in the CDMA IS95 standard [2]. Those methods are used to ensure robust system performance in the presence of typical channel impairments and are outlined next.

5.1.1 Synchronization

Ideally, the digital modulator synchronizes the reverse-link frames with the PN sequences and frame intervals generated at the base station. In practice, that is nearly impossible because of radio propagation effects, so an effective alternative is needed. That alternative is to align the transmit data to the signal

---

1. Synchronization within a few chips.

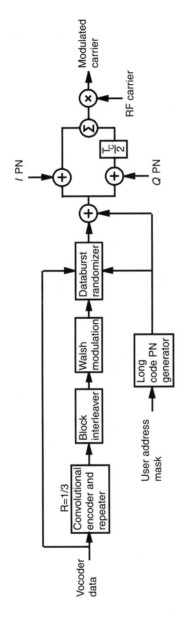

Figure 5.2 Block diagram of digital modulator for CDMA IS95 mobile radio.

received by the mobile radio. That offsets synchronization by the round-trip delay from the base station to the mobile radio and back, which is tolerable.

The time-tracking loop, shown in Figure 5.3(a), maintains system synchronization. It aligns the start of each transmitted frame, as illustrated in Figure 5.3(b). It uses the system timing function to detect the beginning of each received frame and offsets the trigger signal (t_{Tr}) by an amount equal to the processing delay of the digital modem, that is,

$$t_{Tr} = t_{Rx} + nT - (\tau_{Demod} + \tau_{Mod}) \qquad (5.1)$$

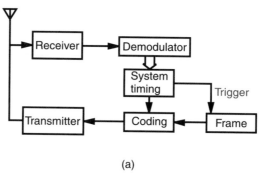

(a)

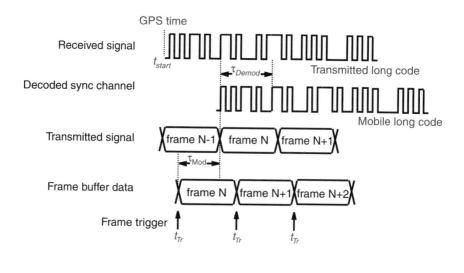

(b)

Figure 5.3 Synchronization: (a) time-tracking loop and (b) timing diagram.

where t_{Rx} is the start of an arbitrary frame, n is an integer, T is the length of a data frame and is equal to 20 ms, τ_{Demod} is the demodulator delay, and τ_{Mod} is the modulator delay. Note that the delay through the RF system is considered negligible.

The synchronization process is simplified because the frame interval is fixed and the entire CDMA IS95 network is based on a common time reference, the GPS [3, 4]. The time reference is communicated via the long-code PN sequence and the forward-link sync channel.

5.1.2 Channel Coding

A key benefit of digital communications is the ability to protect data against channel impairments. The protection is introduced by channel coding. Essentially, the coding adds redundancy, helps identify errors, and provides a way to correct corrupted data. Channel coding is different from source coding, which merely tries to compact the digitized data. Channel coding is implemented in the digital modulator.

The frame buffer of the modulator receives a packet of data prior to each trigger signal. The data packet is coded with a convolutional encoder, a form of digital linear filter that introduces redundancy to the original data sequence and thus provides forward error protection against additive noise in the channel [5, 6].

A simple convolutional encoder is illustrated in Figure 5.4. It consists of two memory devices, two summers, and a multiplexer that operates at twice the original data rate. The output of the encoder is described by

$$L_1 = a(n) + a(n-2) \quad L_2 = a(n) + a(n-1) + a(n-2) \quad (5.2)$$

where a is the input sequence. The multiplexer alternately selects between L_1 and L_2 at the normal clock rate, generates two possible outputs for each input,

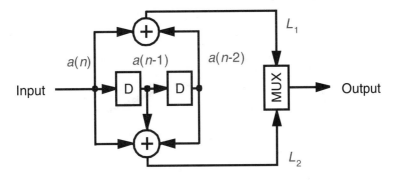

Figure 5.4 Simple rate = 1/2, length = 3 convolutional encoder.

and thus doubles the data rate. The convolutional encoder is characterized by its constraint length and code rate. The constraint length (k) refers to the span of the input sequence processed by the encoder and equals one more than the number of memory devices. The code rate (r) describes the relationship of input bits to output bits. The convolutional code in this simple example is described by ($k = 3$ and $r = 1/2$), while the convolutional code used by the reverse link modulator in a CDMA IS95 mobile radio is specified as ($k = 9$ and $r = 1/3$).

The encoded data is repeated as needed[2] and written by columns into the 18-column by 32-row matrix shown in Figure 5.5. The data is then held until the interleaving process is triggered. This process reads out the data by rows and effectively shuffles the data sequence. Interleaving improves performance for rapidly changing radio channels by introducing time diversity, but it lowers performance in slow-changing radio environments [5]. In practice, the interleaving span is limited to 20 ms—the length of one frame—because longer delays affect voice quality.

The rigid matrix structure of the interleaver produces subframes, known as power control groups, that are 1.25 ms long and are duplicated at data rates less than full-rate. That is, at half-rate, there are eight different subframes that are each repeated two times. At quarter-rate, there are four different subframes that are each repeated four times. And at eighth-rate, there are two different subframes that are each repeated eight times.

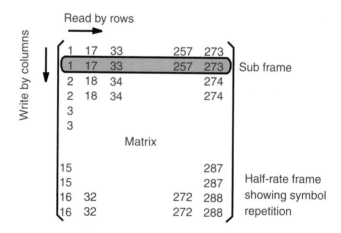

Figure 5.5 The interleaver shuffles the data sequence and thereby improves performance in time-varying channels.

---

2. Data repeats two, four, or eight times, depending on the vocoder data rate.

The output of the interleaver is modulated using Walsh functions. Walsh functions map symbols, six at a time, to one of 64 unique Walsh codes from the Hadamard matrix.[3] The process is not used for orthogonal spreading and is referred to generally as 64-ary modulation or specifically as Walsh modulation.

The Walsh modulated data is then scrambled and spread by the ESN-masked long code. The long code is a PN sequence of 2^{42} chips that repeats every 41 days. It tracks network time and provides a signal to synchronize the mobile radio. The masked-ESN long code is generated by the mobile radio and is offset from the network PN sequence by the ESN of the user. As such, it provides a large number of potential codes for multiple access on the reverse link and scrambles data for added privacy.

The randomizer reduces the average ensemble power of the transmitter. It blanks out redundant power control groups that were generated by the symbol repeater (at vocoder rates of one-half, one-fourth, or one-eighth). The randomizer uses an algorithm based on the long code to pseudorandomly blank the extra power control groups produced by the symbol repeater. That reduces interference, increases system capacity, and improves the bit energy per noise density ratio (E_b/N_o), as shown

$$\frac{E_b}{N_o} \approx \frac{\dfrac{W}{R}}{\nu(1 + f)k} \tag{5.3}$$

where ν is the voice activity rate (typically 3/8 for English speech), f is a factor assigned to "other-cell" interference, and k is the number of users.[4] The randomizer also extends the battery lifetime of the mobile, because the radio transmitter is turned off, or "punctured," when the data is blanked.

The randomized data is then split and covered by I and Q short PN codes. The short codes are distinct 2^{15} chip sequences that are aligned to the forward link pilot by the time-tracking loop. To prevent simultaneous I- and Q-data changes, the Q-data are delayed by one-half chip. That produces offset-QPSK (OQPSK) modulation, reduces amplitude changes in the carrier envelope (because, at most, one bit transition occurs at any time), and relaxes radio circuit design (as is shown in Chapter 8).

The PN sequences used in the modem typically are generated by a maximum-length shift register (MLSR), which is illustrated in Figure 5.6 [7]. It produces a PN sequence that appears to be random but actually repeats

---

3. The six symbols index different rows of the 64-by-64 Hadamard matrix.
4. This expression is derived from (2.20).

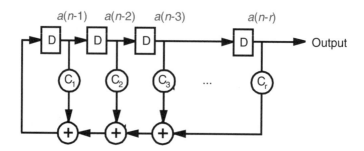

Figure 5.6 MLSR PN generator.

every $2^r - 1$ clock cycles, with r being the number of delay elements in the shift register. The MLSR sequence is described by the linear recursive equation

$$a(n) = c_1 a(n - 1) + c_2 a(n - 2) + \ldots c_r a(n - r) = \sum_{i=1}^{r} c_i a(n - i)$$

(5.4)

where the connection variable, c_i, is either 0 or 1, and addition is modular-2.

It is often useful to shift the PN sequence in time, a process needed to acquire system timing or to produce multiple-access codes. The shift is possible with the masking operation illustrated in Figure 5.7. In this example, the mask {111} delays the PN sequence by two clock cycles.

Note that the MLSR generates an odd number of states and thus an uneven number of logic 0s and 1s. To balance the PN sequence, an extra 1 is added to the end of the sequence.

5.1.3 Signal Filtering

An important figure of merit for wireless communication systems is bandwidth efficiency. It measures the bandwidth occupied by the transmitted signal normalized to the data rate of the message signal. In practice, the message signal is often filtered or pulse-shaped before modulation to contain the spectrum of the transmitted signal, as shown next for some popular modulation schemes.

A BPSK-modulated signal can be described by

$$s(t) = A \cos[2\pi ft + \theta(t)]$$

(5.5)

where $\theta(t) = 0$ when the message data $d(t) = 0$ or π when $d(t) = 1$. Note that (5.5) can be rewritten as

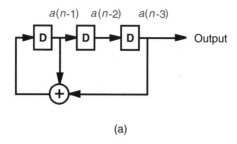

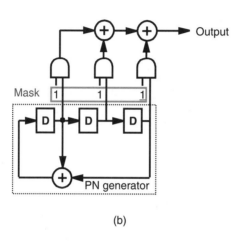

Figure 5.7 Masking operation for PN sequence: (a) three-stage PN sequence shift generator and (b) masking the output delays the sequence two clock cycles.

$$s(t) = Ad(t)\cos 2\pi ft \qquad (5.6)$$

where $d(t)$ is the message data, constructed of rectangular pulses with bipolar values $(-1, +1)$. As a result, the psd of the BPSK-modulated signal is simply the psd of the message data[5] [8], in this case the rectangular or Nyquist pulses.

The psd of the message data is found by first taking the Fourier transform of the signal and then squaring the result. The Fourier transform of the signal $d(t)$, over the bit interval $-T_b/2 < t < T_b/2$, is

$$D(f) = \int_{-T_b/2}^{T_b/2} d(t)e^{-j2\pi ft}dt \qquad (5.7)$$

---

5. Because the cos $2\pi ft$ term provides only frequency translation.

has a psd equal to

$$P_d(f) = \frac{1}{T_b} |D(f)|^2 \tag{5.8}$$

and is shown in Figure 5.8. Furthermore, because $A = \sqrt{E_b/2}$, the psd of the BPSK-modulated signal is simply

$$P_{BPSK}(f) = \frac{E_b}{2} T_b \left(\frac{\sin \pi f T_b}{\pi f T_b} \right)^2 \tag{5.9}$$

A QPSK-modulated signal carries two message bits per symbol using orthogonal BPSK signals, where

$$s(t) = \frac{A}{\sqrt{2}} [d_I(t)\cos\omega t + d_Q(t)\sin\omega t] \tag{5.10}$$

Because each symbol represents two data bits, the symbol period T_s extends to twice the bit period T_b. Note that the psd's of the two orthogonal BPSK signals are identical; therefore, the overall psd is simply

$$P_{QPSK}(f) = E_s T_s \left(\frac{\sin \pi f T_s}{\pi f T_s} \right)^2 \tag{5.11}$$

because $A = \sqrt{2E_s}$ and $E_s = 2E_b$. The striking result of QPSK modulation is that it is two times more bandwidth efficient than BPSK modulation. In addition, the psd of QPSK and OQPSK signals is identical.

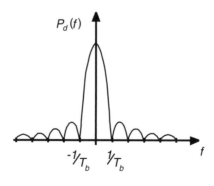

Figure 5.8 The psd of the BPSK-modulated signal is simply the psd of the message data.

In practice, the message data is pulse-shaped to minimize side lobe energy. One example of such an approach is minimum shift keying (MSK). Classical MSK shapes the rectangular data pulses such that

$$d_I(t) \rightarrow d_I(t) \sin\left(\frac{\pi t}{2T_b}\right) \quad d_Q(t) \rightarrow d_Q(t) \sin\left(\frac{\pi t}{2T_b}\right) \tag{5.12}$$

and thereby avoids phase discontinuities at the beginning and the end of the data pulses [8]. It has a psd described by

$$P_{MSK}(f) = \frac{8E_b}{\pi^2} T_s \left(\frac{\cos \pi f T_s}{1 - (4fT_s)^2}\right)^2 \tag{5.13}$$

where the main lobe is extended to $1.5/T_s$.

MSK modulation is a type of constant envelope modulation. Constant envelope modulation schemes provide the following advantages [9]:

- Extremely low side-lobe energy;
- Use of power-efficient class-C or higher amplifiers;
- Easy carrier recovery for coherent demodulation;
- High immunity to signal fluctuations.

A variant of MSK modulation is GMSK modulation. It shapes the message data with a filter that further reduces side-lobe energy [9, 10]. The impulse response of the filter is described by

$$h_{GMSK}(t) = \frac{\sqrt{\pi}}{\alpha} \exp\left(-\frac{\pi^2}{\alpha^2}t^2\right) \tag{5.14}$$

where $\alpha = 0.5887/B$.[6] Note that when $B = 0.5887$, the side lobes of the modulated signal virtually disappear.

The psd's of these modulation schemes are plotted in Figure 5.9. The plot shows that linear modulation schemes, such as BPSK, QPSK, and OQPSK, have a null at $1/T_s$ with higher side lobe energy. In contrast, the psd's of signals generated by constant-envelope modulation techniques, like MSK and GMSK, have a wider main lobe with lower side-lobe energy, due primarily to filters

---

6. Modulation filters generally are defined by the product BT, where T is the symbol rate.

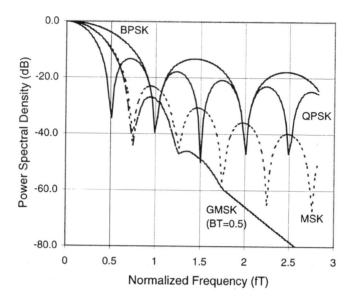

Figure 5.9 A comparison of the signal bandwidths for some popular modulation schemes.

that reduce the phase discontinuities at the beginning and the end of the data pulses [11].

For optimal performance, the modulation filter should shape each pulse in the data sequence such that the overall response of the communication system (transmitter, channel, and receiver) at any given sampling instant is zero, except for the current symbol, as depicted in Figure 5.10. That, in effect, nulls the interference between symbol pulses, a condition known as the Nyquist criterion for intersymbol cancellation [12].

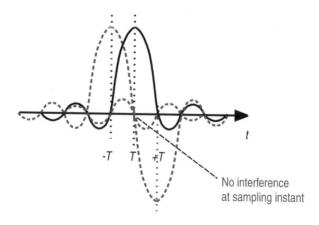

Figure 5.10 Nyquist criterion for eliminating intersymbol interference.

The most popular pulse-shaping filter for wireless communications is the raised cosine filter. It has a transfer function given by

$$H(f) = \begin{cases} = 1 & 0 < f \le \dfrac{(1 - \alpha)}{2T} \\[12pt] \dfrac{1}{2}\left[1 + \cos\left(\pi\dfrac{2Tf - 1 + \alpha}{2\alpha}\right)\right] & \dfrac{(1 - \alpha)}{2T} < f \le \dfrac{(1 + \alpha)}{2T} \\[12pt] = 0 & f > \dfrac{(1 + \alpha)}{2T} \end{cases} \qquad (5.15)$$

where α is the bandwidth expansion factor. It is given that name because of its effect on the main lobe of the modulation signal's psd, as shown in Figure 5.11. In practice, the raised cosine filter typically is split between the transmitter and the receiver into two root raised cosine filters equal to $\sqrt{[H(f)]}$.

The pulse-shaping filter used in CDMA IS95 communication systems is extremely narrow. It is a 48-tap symmetric FIR structure, with linear phase response, low in-band ripple (less than 1.5-dB variation from dc to 590 kHz), and high out-of-band attenuation (greater than 40 dB at 740 kHz) [13]. Consequently, the spectrum of the filtered CDMA IS95 waveform is contained to the main lobe, as shown in Figure 5.12. Unfortunately, the filter is not a Nyquist filter and thus creates overshoot in the time domain and introduces intersymbol interference (ISI).

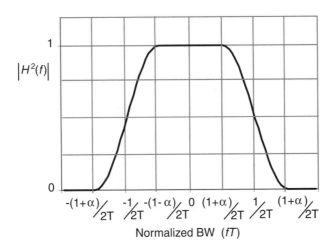

Figure 5.11 Transfer function of the raised cosine filter.

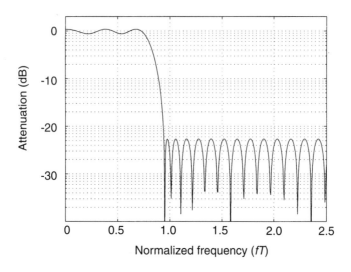

Figure 5.12 Spectrum of CDMA IS95 signal after filtering.

5.2 Digital Demodulator

The most complicated function in the digital system is the digital demodulator. It is responsible for recovering the transmitted message signal after the wireless channel has distorted it. That formidable task directly affects the performance of the mobile radio's receiver.

The digital demodulator consists of the searcher, the Rake receiver, and other digital signal processing functions, as shown in Figure 5.13. The searcher synchronizes the mobile radio's internal PN generators to the received pilot

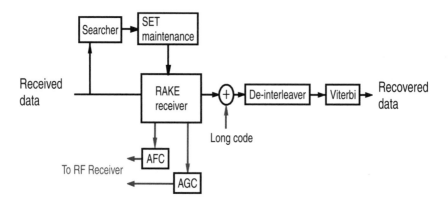

Figure 5.13 Block diagram of digital demodulator for CDMA IS95 mobile radio.

channel, a process known as pilot acquisition. The Rake receiver then uses this phase reference for coherent detection of the received data. To recover the transmitted data, the digital demodulator typically decodes the received symbols using the Viterbi algorithm.

Typically, two feedback control systems are used in the digital demodulator to track the strength and carrier frequency of the received signal. The AFC loop corrects the RF synthesizer to achieve perfect baseband signals after downconversion. The AGC loop adjusts the gain of the radio receiver to overcome fading effects introduced by the wireless channel.

5.2.1 Pilot Acquisition

The first task of the digital demodulator is pilot acquisition. This process analyzes the signals received by the radio receiver, including a wide spectrum of interference and noise plus several CDMA channels[7] at the selected radio frequency. Ideally, the radio receiver attenuates the interfering signals and leaves only the signals at the selected carrier frequency, corresponding to the different CDMA channels and their associated multipath components.

All forward-link transmissions share one important characteristic, a dominant pilot channel, which is just the short PN (2^{15} chips) sequence signature. The pilot acquisition function employs a searcher algorithm [14], which correlates the input data against internally generated I and Q PN sequences using

$$R_i = \sum_N r(t)pn_i(t) \tag{5.16}$$

where N is the cross-correlation length, $r(t)$ is the received signal, and pn_i is the ith offset of the PN sequence. (Note that the offset is formed by a masking operation.) Each possible offset of the short PN sequence must be tested to identify the strongest pilot signals and ensure acquisition. That means that if the digital receiver has a resolution of one-fourth chip ($T_c/4$) there are 2^{15+4} test hypotheses.

To accelerate the searching process, double-dwell algorithms [15] typically are used. An initial correlation of L_1 samples (where $L_1 < N$) is computed and compared to a threshold θ_1. If it fails, the next hypothesis is checked. If successful, the dwell is increased to L_2 samples, and the correlation result is compared to θ_2. If it succeeds, the hypothesis is considered correct, and the PN offset is forwarded to the set maintenance block. If the second test is unsuccessful, the hypothesis is discarded. By quickly eliminating unlikely

---

7. CDMA channels encompass the pilot, sync, and paging channels, plus multiple traffic channels.

hypotheses with a short initial correlation (of L_1 samples), the overall acquisition time is reduced. To further reduce acquisition time, parallel correlators can be used.

The set maintenance function organizes the results of the searcher algorithm using information provided by the network over the paging channel. The information denotes the strong pilot signals by PN offset and, in general, classifies the PN offsets into one of four categories:

- The *active set* states the PN offsets of the base stations transmitting valid signals to the mobile radio.
- The *candidate set* lists the PN offsets that the mobile radio considers strong enough for the active set.
- The *neighbor set* includes the PN offsets of nearby base stations.
- The *remaining set* captures the weaker PN offsets.

In practice, the mobile radio is typically in soft handoff and is receiving transmissions from two or three different base stations. The set maintenance function recognizes those active PN offsets and their multipath components and forwards the PN offsets with the highest cross-correlation results to the Rake receiver.

It is crucial to accurately identify the timing of the short PN sequence. That is because the despreading process is implemented by a simple correlator described by

$$\frac{1}{T_b} \int_0^{T_b} r(t)pn(t)dt \qquad (5.17)$$

where T_b is the bit period of the message signal. The result is essentially the cross-correlation between the received PN sequence and the internally generated PN sequence aligned by the pilot acquisition process. The autocorrelation of the PN sequence is very small for offsets (τ) greater than the period of a chip (T_c) and can be approximated by a piecewise linear function [14], with values

$$R(\tau) = 1 - \frac{|\tau|}{T_c} \qquad |\tau| \le T_c \qquad (5.18)$$

Hence, the output of the demodulator is proportional to $R(\tau)$ and is very small when there are bit synchronization errors.

5.2.2   Carrier Recovery

The carrier recovery loop links the digital demodulator to the radio receiver. It employs feedback to phase-lock the radio receiver to the transmitted carrier frequency, as shown in Figure 5.14. This minimizes phase errors in the data detection process, an important consideration in phase-modulated systems.

The detection process maps samples of the received signal to the complex plane. Ideally, the samples occur at distinct modulation points and form the constellation diagram shown in Figure 5.15. In practice, the samples follow the trajectory of the received signal and scatter when the timing of the transmitter and the receiver differ.

The radio receiver downconverts the received signal to baseband. Ideally, it translates the carrier frequency to dc and thus aligns the transmitter and the receiver. Any frequency error, in effect, rotates the data samples about the complex plane at the error frequency and causes detection errors.

Coherent detectors rely on a reference signal to align the receiver to the transmitter. CDMA IS95 mobile radios use the pilot signal as that reference signal. It is chosen because the pilot signal is a relatively strong signal that is aligned to the sync, paging, and traffic channels and is subjected to the same radio propagation effects.

To assess carrier recovery in the radio receiver, the demodulator tracks the phase of the pilot signal. This is rather straightforward because the transmitted data for this channel is an all-zero sequence. As a result, the phase of the pilot signal is found by using the magnitudes of the I and Q components of the received signal and the simple trigonometric relation:

$$\theta = \tan^{-1}\left(\frac{I}{Q}\right) \tag{5.19}$$

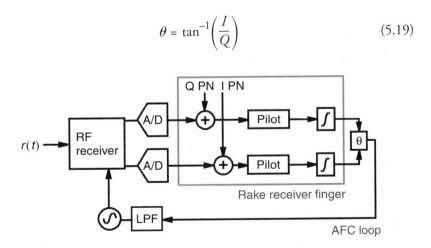

Figure 5.14 AFC loop for carrier recovery.

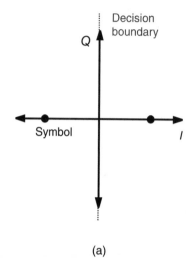

(a)

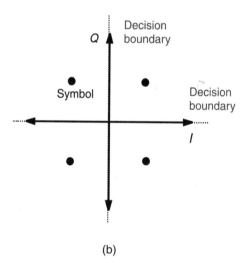

(b)

Figure 5.15 Constellation diagrams: (a) BPSK signals and (b) QPSK signals.

From (5.19), the frequency difference between the transmitted carrier and the radio receiver is found using

$$\omega_\epsilon = \frac{1}{T^2} \int_T \theta(t)\,dt \qquad (5.20)$$

The result is fed back to the RF synthesizer, where minor adjustments are made. Note that large frequency errors cause the detected data to jump around and make it difficult to analyze the trajectory of the pilot signal. Such errors generally are handled by an FFT algorithm.

The impact on receiver performance for small, bounded phase errors is analyzed by determining the effect on received bit energy. The received direct-sequence spread-spectrum signal $r(t)$ is described by $pn(t)Ad(t)\cos\omega t$. After downconverting, filtering, and despreading, the received signal is transformed to

$$E_b(t) \propto Ad(t)\cos(\omega_\epsilon t) \tag{5.21}$$

where $E_b(t)$ is the energy per bit of the received signal, ω_ϵ is due to the carrier synchronization error, and the product $\omega_\epsilon t$ is the instantaneous phase error θ_ϵ. That means the amplitude of the received bit energy decreases with $\cos(\theta_\epsilon)$. It also means the probability of detection error increases as the samples move closer to the decision boundaries, as shown in Figure 5.16. For BPSK-modulated data, the probability of error is given by (3.32), which can be augmented for phase error as shown by [16]

$$P_e = Q\left[\sqrt{\frac{2E_b}{N_o}}\cos(\theta_\epsilon)\right] \tag{5.22}$$

where $Q[*]$ is the complimentary error function, and θ_ϵ is the root-mean-square (rms) phase error. For QPSK-modulated data [16], the probability of error expands to

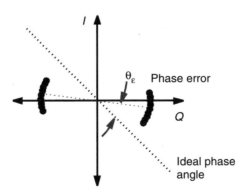

Figure 5.16 Effect of phase error on the constellation diagram.

$$^2[\cos(\theta_\epsilon) + \sin(\theta_\epsilon)]\bigg]_o + \frac{1}{2}Q\bigg[\sqrt{\frac{2E_b}{N_o}}[\cos(\theta_\epsilon) - \sin(\theta_\epsilon)]\bigg]$$

$$(5.23)$$

wi.. the leakage of orthogonal signal components as the frequency error rota.. the data in the complex plane.

5.2.3 Signal Leveling

The AGC loop provides a second link from the digital demodulator to the radio receiver. It uses the feedback loop shown in Figure 5.17 to maintain a relatively constant signal level at the input to the A/D converters.[8] The task is challenging because the received signals are affected by large-scale attenuation and multipath fading introduced by the wireless channel. A typical received signal, shown in Figure 5.18, is characterized by rapid level changes. The increases in power level are known as upfades and generally are limited to about 6 dB above the rms level [17]. The decreases in power level are known as downfades and are typically sharp and occasionally dramatic.

Practical A/D converters are sensitive to a limited range of input levels. These circuits have a fixed noise floor and thus their performance (SNR)

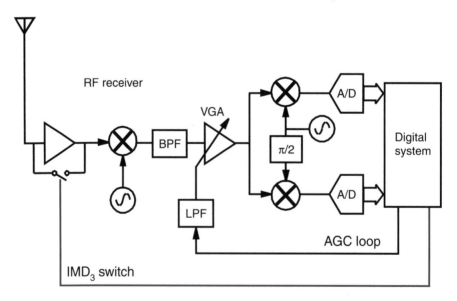

Figure 5.17 The AGC loop strives to maintain a relatively constant voltage level to the A/D converters.

8. Two A/D converters are used to translate the I and Q signals to digital format.

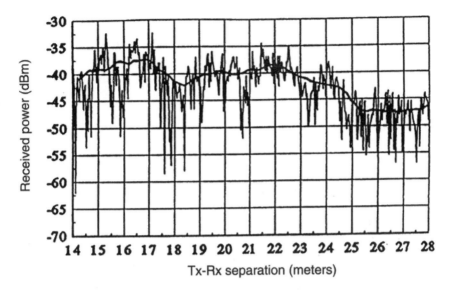

Figure 5.18 Plot of received power for a mobile radio (*From:* T. S. Rappaport, *Wireless Communications*, © 1995, reprinted by permission of Prentice Hall Inc., Upper Saddle River, NJ).

degrades at lower input levels. Consequently, to achieve optimum performance, the rms value of the input signal typically is centered at approximately 6 dB below the full-scale value of the A/D converter.[9] In fact, that is the objective of the AGC algorithm.

The AGC algorithm is based on the rms level of the received signal, which is defined as

$$V_{RMS} = \sqrt{\frac{1}{T} \int_T V^2(t)\,dt} \tag{5.24}$$

and is a measure of the average power over the interval T [17]. It is convenient to rewrite (5.24) expression as

$$V_{RMS} \approx \sqrt{\frac{1}{N} \sum_N V^2(t)} \tag{5.25}$$

---

9. The actual rms value depends on the input range of the A/D converter.

for digital systems, where N is the equivalent number of samples. In either case, the rms expression relies on the square root function. In general, that function is not readily available in DSP hardware and is therefore inefficient. As such, an approximation like the logarithm function is oftentimes preferable, where

$$V_{RMS}(dB) \propto \log\left(\frac{1}{\sqrt{N}}\right) + \frac{1}{2}\log\left[\sum_N (I^2 + Q^2)\right] \qquad (5.26)$$

which simplifies to

$$V_{RMS}(dB) \propto \log\left[\sum_N I^2 + \sum_N Q^2\right] \qquad (5.27)$$

Note that the approximation expresses the signal power in decibels and thus provides the benefit of compactly describing the wide range of received signal levels.

The AGC algorithm for closed-loop and open-loop power control is shown in Figure 5.19. It includes a digital FIR filter to stabilize the loop and ensure that the system tracks the average rms level instead of received signal fluctuations. In the CDMA IS95 mobile radio, the computed rms value for

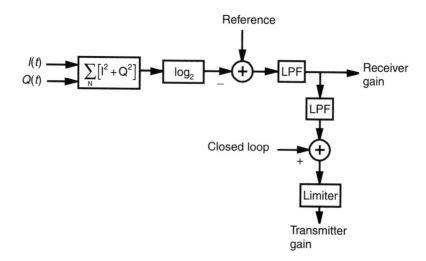

Figure 5.19 Block diagram of the AGC algorithm used to maintain A/D converter input levels and to set the open-loop transmit power level.

the received signal feeds the open-loop power control network that sets the transmit power level. It is based on the expression[10]

$$P_{Tx} = -73dBm - P_{Rx} \qquad (5.28)$$

where P_{Rx} is the received power level and P_{Tx} is the transmit power level. The open-loop response of the transmitter is set much slower than the receiver AGC loop. That prevents the transmitter from following the sharp downfades of the received signal.

The AGC algorithm also includes closed-loop logic that detects power control information sent by the base station to offset the transmitter AGC loop. The information is extracted from each received subframe or power control group at a data rate of 800 Hz.

The AGC algorithm also monitors the frequency spectrum of the received waveform. The spread-spectrum waveform is nominally flat, but strong interferers produce intermodulation products "in-band" that degrade the detection process. Those products are easily distinguished because they appear relatively narrow in the FFT output. In such cases, the front-end gain is reduced to minimize distortion in later stages through the use of a switch around the LNA, as shown in Figure 5.17.

5.2.4 Data Detection

Direct-sequence spread-spectrum communication systems utilize the Rake receiver (an extension of the matched correlator receiver) for data detection. The Rake receiver consists of parallel correlators known as fingers and a maximal ratio combiner, as shown in Figure 5.20. The correlators are set up to resolve the strongest multipath signals arriving at the receiver [14, 15]. The signals are identified by the searcher algorithm and are specified by relative offsets in the short PN sequence. As such, the correlator function can be written as

$$z(T) = \int_{\tau}^{T+\tau} r(t)pn(t - \tau)dt \qquad (5.29)$$

where τ is the normalized multipath delay.

The maximal ratio combiner sums the output of the matched correlators and thereby increases the aggregate signal power [18]. The combiner's output is

---

10. Cellular band expression is shown; for PCS band, the offset parameter is −76 dBm.

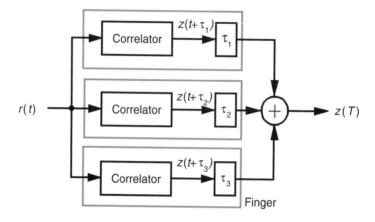

Figure 5.20 Block diagram of a Rake receiver.

$$z_k(T) = \sum_k \int_{\tau_k}^{T+\tau_k} r(t)pn(t - \tau_k)dt \qquad (5.30)$$

where k is the number of fingers in the Rake receiver, typically between three and six, and τ_k is the excess delay associated with each of the dominant multipath components. Essentially, the Rake receiver implements the approach suggested by (1.5) to mitigate the effects of multipath fading.

The operation of the Rake receiver finger, shown in Figure 5.21, is key to the data detection process. It isolates one of the strong multipath components, provides bit synchronization, detects the pilot data and notes rotation, estimates

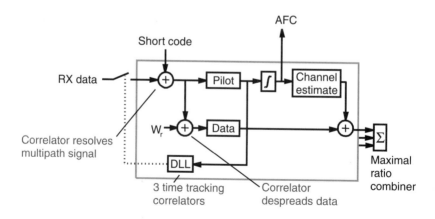

Figure 5.21 Rake receiver finger.

the amplitude and phase characteristics of the radio channel, and despreads the message data.

The finger uses a correlator and the assigned PN sequence to isolate the designated multipath component. Conceptually, the correlator resolves the multipath component and attenuates any other signals.

In CDMA IS95 communication systems, the base station modulator provides the same data to both the I-channel and the Q-channel. As such, those channels can be combined after the short PN correlators to double the signal energy.

Each finger contains four additional correlators: three dedicated to timing recovery and one reserved for data demodulation. The three time-tracking correlators maintain bit sychronization. The correlators operate on different sampling phases of the received data stream. The sampling phases are the result of oversampling the received data stream and are typically spaced one-half chip ($T_c/2$) apart.

The three time-tracking correlators are labeled *early, on time,* and *late.* By design, the on-time correlator matches the data correlator, while the others operate one-half chip before and after. To assess the sampling performance, the autocorrelation for each timing phase is computed using

$$R(0) = \sum_N r^2(n) \qquad (5.31)$$

where $R(0)$ is the average power of the pilot signal $r(n)$, and N is the number of samples.

The time-tracking correlators feed an algorithm that centers the data detection process. The autocorrelation of the early and late samples, $nT - \Delta$ and $nT + \Delta$, respectively, are

$$R_- = R(-\Delta) \qquad R_+ = R(+\Delta) \qquad (5.32)$$

where $\Delta = T_c/2$. Ideally, the sample times lie on opposite sides of the autocorrelation main lobe and $R_- - R_+ \to 0$, as shown in Figure 5.22. If the timing is early, R_- will be smaller than R_+. Conversely, if the timing is late, R_- will be larger than R_+. The time-tracking algorithm assesses bit sychronization using the following formula

$$Error = R^2(-\Delta) - R^2(+\Delta) \qquad (5.33)$$

and strives to minimize the error by way of the delay-locked loop (DLL), which advances or retards the sampling phase to keep the error minimized [19].

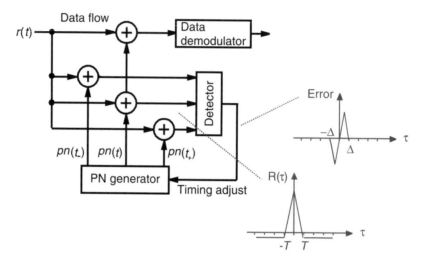

Figure 5.22 Block diagram of the delay-locked loop and early-late time tracking process.

The on-time correlator also provides the data used for carrier recovery. Note that the output does not need to be despread or decoded because the pilot channel is formed from an all-zero data sequence and the all-zero Walsh function.

A single correlator in the finger is reserved for data demodulation and spans 64 chips, the equivalent of one symbol period

$$z(T) = \int_{\tau}^{T+\tau} r(t)pn(n - \tau)w_r(t)dt \qquad (5.34)$$

where τ is the PN sequence delay and $w_r(t)$ is the Walsh code assignment. A processing gain of 64x (18 dB) is realized by this correlator and despreading operation.

Each finger outputs soft finger data, $z(T)$, which is deskewed and scaled according to the finger's assigned index value and the strength of the multipath component. This process is known as channel estimation and is partially accomplished by the data correlator.[11] The maximal ratio combiner constructively adds the outputs to produce soft Rake receiver data, $\bar{z}(T)$, given by

---

11. The amplitude of the multipath component and hence the scaling is preserved by the correlator.

$$\overline{z}(T) = \sum_k z_k(T - \tau_k) \qquad (5.35)$$

where $z_k(T - \tau_k)$ is the soft finger decision given by (5.34).

5.2.5 Data Recovery

The detection process removes the modulation from the received signal but does not recover the message data. That is because the message data is still protected by convolutional coding, block interleaving, and scrambling applied at the base station transmitter. The recovery process translates the data produced by the Rake receiver to an estimate of the original message.

The data from the Rake receiver, $\overline{z}(T)$, is first unscrambled. This operation removes the long code added by the forward-link modulator (described in Chapter 2). It requires synchronization of an internal PN generator to the sequence received by the mobile radio, a rather straightforward task since the base station transmits the value of the long-code generator advanced by 320 ms.

After unscrambling, the recovery process deinterleaves the data. This operation reverses the interleaving operation performed by the forward-link modulator.

Last, the recovery process decodes the data. This is a challenging task because the forward-link modulator first encodes the message data and then repeats the encoded bits as needed to fill the data frame. (Recall that the number of message bits varies with the variable rate of the vocoder.) To complicate matters further, the base station does not transmit the vocoder data rate and thus requires rate determination by the mobile radio demodulator. As a result, all four data rates are demodulated and their results are verified against the CRC.[12]

To illustrate the Viterbi decoding process, the simple convolutional encoder shown in Figure 5.4 is used. The operation of the convolutional encoder can be conveniently described by the trellis diagram [20, 21] shown in Figure 5.23. In the trellis diagram, the shift register's contents are shown as nodes, the next bit of the input sequence is shown as a branch, and the output of the convolutional encoder is shown next to each branch. For this example, each output symbol is two bits, because the encoder is rate one-half.

The Viterbi algorithm [22] is a maximum likelihood detector. It uses the received data to reconstruct the transmitted data sequence and thus the message signal. The challenge comes from noise and other impairments that

---

12. The CRC is available only at full and half rates for rate set 1 but at all rates for rate set 2 (see Chapter 2 for information about rate sets).

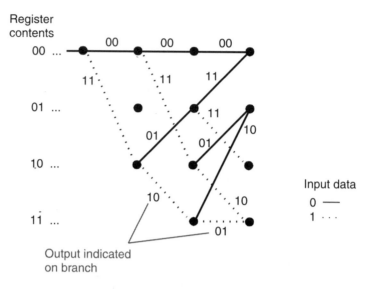

Figure 5.23 Trellis diagram for convolutional encoder shown in Figure 5.4.

alter the received data. To combat those effects, convolutional codes as well as block codes [20] use redundancy.

As a maximum likelihood detector, the Viterbi algorithm analyzes the conditional probabilities for each branch in the trellis diagram given the received data. It does that over a sequence of data to account for the depth or constraint length of the encoder. Furthermore, the Viterbi algorithm exploits a given observation to simplify the decoding routine [23].

The conditional probability $p(y|x)$ describes the likelihood that the transmitted data sequence or vector x was sent given the received data vector y. This function computes the euclidean distance between the received data and defined symbol states shown in the trellis diagram. The conditional probability for a path through the trellis diagram is given by

$$p(y|x) = \prod_n p[y(n)|x(n)] \tag{5.36}$$

where n indexes all successive symbols associated with the branches that form the path. Taking the logarithm of both sides of (5.36) greatly simplifies its computation and yields the simple, additive function

$$\Lambda(y|x) = \ln p(y|x) = \sum_{all\ n} \ln p[y(n)|x(n)] \tag{5.37}$$

The term $\ln p[y(n)|x(n)]$ is related to the branch metric $m(n)$ as shown:

$$m(n) = \mu \ln p\,[y(n)|x(n)] \qquad (5.38)$$

where μ is a proportionality constant. In general, the maximum likelihood detector searches for the data sequence, or path, through the trellis diagram where the sum of the branch metrics is maximized.

As described, the maximum likelihood detector is computationally cumbersome because the number of paths doubles at each state. However, that can be simplified by noting that when two paths meet at a node the path with higher metric is the solely important path. All other paths can be discarded. As a result, the number of traced paths equals, at most, the number of nodes in the trellis diagram.

An illustration of the Viterbi algorithm applied to the example convolutional encoder is shown in Figure 5.24. It uses a simple branch metric that counts the number of matching bits between the received data and the synthesized sequence found from the trellis diagram.

The tail bits are a key part of the decoding process. They "flush out," or clear, the convolutional encoder and provide a known starting point for the decoder. As such, the input and the contents of the convolutional encoder at the end of the data frame are known. Note that the Viterbi algorithm works backward from the last bits to the start of the data frame.

The contents of the encoder are 00 at the end of the data frame; note that there are only two possible branches that connect to this node. From the trellis diagram, the input 0 produces the upper branch and the output 00, while the input 1 produces the lower branch and the output 11. These possible output codes are compared to the received data, and branch metrics are computed, as shown in Figure 5.24(a).

At this point, there are two possible nodes, 00 and 01. In addition, the trellis diagram shows that each node supports two branches. The decoding algorithm obtains the expected output code for each branch, compares those codes to the received data, and computes the associated metrics. The results are shown in Figure 5.24(b).

The process continues backward until the entire received frame has been analyzed. Note that after three symbols the paths remerge and the lower ranking paths get discarded, as shown in Figures 5.24(c) and 5.24(d).

At half-rate and full-rate, the CRC verifies the original message data. At lower rates, the additional redundancy due to the symbol repetition improves data recovery. However, if the data frame is analyzed without success, it is discarded and a frame erasure is reported.

The Viterbi algorithm is executed with dedicated add-compare-select (ACS) hardware [7]. Its operation is as follows: *Add* each branch metric to the preceding level for the allowable transitions; *compare* the pair of metric sums

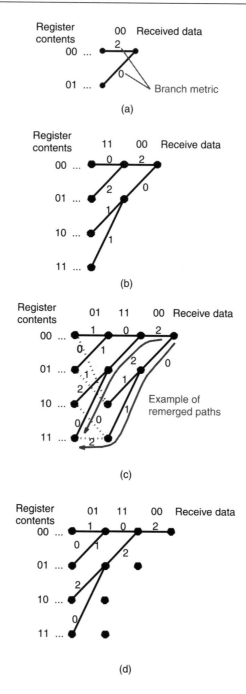

Figure 5.24 Viterbi algorithm: (a) last node branches, (b) last two node branches, (c) last three node branches, and (d) with remerged paths removed.

for paths entering a state node; and *select* the greater of the two paths and discard the other. If two quantities are the same, either branch can be selected, because each has equal probability.

To implement the Viterbi decoder, two sets of results are stored. The first set tracks the metric computations and is updated at each node state. The second set is the data selected at each node state and is ultimately the desired message signal. The final decision is made by a chaining-back procedure, starting with the last decision and moving back to the first. The chaining-back procedure does not have to cover an entire frame, merely the distance between remerged paths. That distance is the traceback length of the algorithm.

The convolutional encoder and Viterbi algorithm work together to provide data protection. The amount of data protection is linked to the structure of the convolutional encoder but limited by Shannon's capacity theorem [24]. Shannon's theorem states that it is possible to transmit information over any channel (with sufficient capacity, C) at a rate R with arbitrarily small error probability by using a sufficiently complicated coding scheme. The capacity of a channel, perturbed by AWGN, is described by

$$C = W \log_2\left(1 + \frac{S}{N}\right) \qquad (5.39)$$

where S is the signal power, N is the noise power, and W is the bandwidth. This expression limits the transmission rate and illustrates the power-versus-bandwidth tradeoff.

Remember, the bit energy per noise density ratio, E_b/N_o, is simply

$$\frac{E_b}{N_o} = \frac{S}{N}\frac{W}{R} \qquad (5.40)$$

where W/R is the spreading rate. Combining (5.39) and (5.40) provides

$$C = W \log_2\left(1 + \frac{E_b}{N_o}\frac{R}{W}\right) \qquad (5.41)$$

and the highest level of protection (as measured by required E_b/N_o).

The performance of the decoding process is also limited by the resolution of the digital hardware. In CDMA IS95 communication systems, the convolutional encoders are constraint length $k = 9$; thus, there are 2^8 different traceback paths. In practice, it is unreasonable to store a large number of bits for each path, so it is necessary to make compromises. These compromises address branch metric computations.

There are two approaches to the computations of branch metrics: euclidean distance [6] and Hamming distance [6]. The euclidean distance is the geometric distance between the possible codes and the received data. Its accuracy is limited by the resolution of the received data. The Hamming distance is computed by first translating the received data to nearest possible code. Although this simplifies computations, it also eliminates any grayness in the received data and lowers the accuracy of the branch metric. Soft-decision algorithms rely on euclidean distances, while hard-decision algorithms use Hamming distances.

The constraint length and the code rate both play an important role in the performance of the convolutional encoder, as shown in Table 5.1. The table shows the benefit of increasing the code rate and using soft decisions. It also should be noted that coding effects are even more dramatic in Rayleigh fading environments.

Table 5.1

Comparison of AWGN Performance for Different Convolutional Codes When the Probability of Bit Error Requirement Equals 10^{-3} [25–27]

Convolutional Code	E_b/N_o Value (dB)	Coding Gain (dB)
No coding	6.8	0
Hard decision		
Rate 1/2, length 5	5.3	1.5
Rate 2/3, length 5	5.6	1.2
Soft decision		
Rate 1/2, length 5	3.2	3.6
Rate 1/2, length 7	2.7	4.1
Rate 1/2, length 9	2.5	4.3
Rate 1/3, length 9	2.2	4.6
Rate 2/3, length 5	3.7	3.1
Rate 2/3, length 7	3.2	3.6
Ideal system		
Shannon's limit	−1.6	8.4

References

[1]   Sklar, B., *Digital Communications*, Englewood Cliffs, NJ: Prentice Hall, 1988.

[2]   TIA/EIA Interim Standard, "Mobile Station-Base Station Compatibility Standard for Dual-Mode Wideband Spread Spectrum Cellular System," IS95a, Apr. 1996.

[3]   Kaplan, E., editor, *Understanding GPS: Principles and Applications*, Norwood, MA: Artech House, 1996.

[4]   Enge, P., and P. Misra, Introduction to *GPS—The Global Positioning System*, special issue of *IEEE Proc.*, Jan. 1999, pp. 3–15.

[5]   Couch, L. W., *Digital and Analog Communication Systems*, Upper Saddle River, NJ: Prentice Hall, 1997.

[6]   Proakis, J. G., *Digital Communications*, New York: McGraw-Hill, 1995.

[7]   Viterbi, A. J., *CDMA: Principles of Spread Spectrum Communications*, Reading, MA: Addison-Wesley, 1995.

[8]   Taub, H., and D. L. Schilling, *Principles of Communication Systems*, New York: McGraw-Hill, 1986.

[9]   Rappaport, T. S., *Wireless Communications: Principles and Practice*, Upper Saddle River, NJ: Prentice Hall, 1996.

[10] Murota, K., and K. Hirade, "GMSK Modulation for Digital Mobile Radio Telephony," *IEEE Trans. on Communications*, Vol. COM-29, No. 7, July 1981.

[11] Murota, P. S., T. L. Singhal, and R. Kapur, "The Choice of a Digital Modulation Scheme in a Mobile Radio System," *Proc. IEEE Vehicular Technology Conf.*, 1993, pp. 1–4.

[12] Nyquist, H., "Certain Factors Affecting Telegraph Speed," *Bell Systems Tech. J.*, Vol. 3, pp. 324–346.

[13] Hinderling, J. K., et al., "CDMA Mobile Station ASIC," *IEEE J. of Solid State Circuits*, Vol. 28, No. 3, Mar. 1993, pp. 253–260.

[14] Peterson, R. L., R. E. Ziemer, and D. E. Borth, *Introduction to Spread Spectrum Communications*, Upper Saddle River, NJ: Prentice Hall, 1995.

[15] Simon, M. K., et al., *Spread Spectrum Communications*, Rockville, MD: Computer Science Press, 1985.

[16] Howland, R. L., "Understanding the Mathematics of Phase Noise," *Microwaves & RF*, Dec. 1993, pp. 97–100.

[17] Van Valkenburg, M. E., *Network Analysis*, Englewood Cliffs, NJ: Prentice Hall, 1974.

[18] Brennan, D. G., "Linear Diversity Combining Techniques," *IRE Proc.*, Vol. 47, 1959, pp. 1075–1102.

[19] Spilker, J. J., Jr., "Delay-Lock Tracking of Binary Signals," *IEEE Trans. on Space Electronics and Telemetry*, Mar. 1963, pp. 1–8.

[20] Lin, S., and D. J Costello, Jr., *Error Control Coding: Fundamentals and Applications*, Englewood Cliffs, NJ: Prentice Hill, 1983.

[21] Lee, C., *Convolutional Coding: Fundamentals and Applications*, Norwood, MA: Artech House, 1997.

[22] Viterbi, A. J., "Error Bounds for ConVolutional Codes and an Asymptotically Optimum Decoding Algorithm," *IEEE Trans. on Information Theory*, IT-13, 1967, pp. 260–269.

[23] Forney, G. D., Jr., "The Viterbi Algorithm," *IEEE Proc.*, Vol. 61, No. 3, Mar. 1973, pp. 268–278.

[24] Shannon, C. E., "Communication in the Presence of Noise," *IRE Proc.*, Vol. 37, 1949, pp. 10–21.

[25] Viterbi, A. J., "Convolutional Codes and Their Performance in Communication Systems," *IEEE Trans. on Communications Technology*, Vol. COM-19, Oct. 1971, pp. 751–772.

[26] Oldenwalder, J. P., "Optimal Decoding of Convolutional Codes," Ph. D. dissertation, UCLA, 1970.

[27] Hardin, T., and S. Gardner, "Accelerating Viterbi Decoder Simulations," *Communications System Design*, Jan. 1999, pp. 52–58.

6

Data Converters

Data conversion is an essential process in digital communication systems because radio and voice signals are naturally analog. These signals interface with the digital system, where source and channel coding/decoding occurs. As such, A/D and D/A converters are needed, as illustrated in Figure 6.1.

The characteristics of the analog signals largely affect the design of the data converters. The radio signals are direct-sequence spread-spectrum modulated and thus are wideband. In contrast, audio signals are narrowband, generally limited to 4 kHz or less, and very dynamic.

This chapter covers the basics of A/D conversion. It identifies the ideal and nonideal distortion mechanisms that plague A/D converters. This leads to a comparison of popular A/D converter architectures that address the nonide-

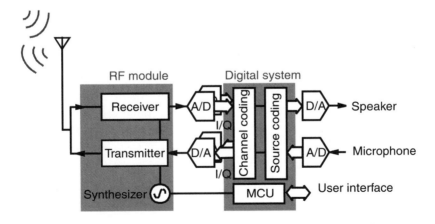

Figure 6.1 A/D interfaces in a typical CDMA mobile radio.

121

alities and target different signal characteristics. The chapter then presents the issues associated with D/A converters and concludes with a review of widely used D/A converter architectures.

6.1 A/D Conversion

The conversion of analog signals to digital form involves two processes: sampling and quantization [1]. The sampling process converts the continuous-time signal to discrete-time samples. The quantization process maps the discrete-time analog samples to digital codes. The ideal quantization process introduces distortion, which grows if the sampling is nonideal or if there are errors in the quantization process.

6.1.1 Ideal Sampling Process

The sampling process takes a "picture" of the analog waveform at discrete points in time, as shown in Figure 6.2. It does that by multiplying the analog waveform $x(t)$ by a train of unit impulse functions $\delta(t - nT)$, that is,

$$y(t) = x(t) \sum_{n=-\infty}^{+\infty} \delta(t - nT) \qquad (6.1)$$

where $y(t)$ are the discrete analog samples and T is the sampling period. Note that the sampled waveform is zero except at integer values of T. Furthermore, the value of T is chosen to meet the Nyquist criterion ($T \leq 1/2B$), where B is the bandwidth of $x(t)$.

The quantization process converts the discrete analog samples $x(t)$ to digital codes $x(n)$, using the transfer function illustrated in Figure 6.3. The quantization process introduces irreversible distortion[1] because each unique digital code represents a range of analog values. That distortion, known as the quantization error, varies between $-\Delta/2$ and $+\Delta/2$. The parameter Δ is defined as

$$\Delta = \frac{V_{max} - V_{min}}{2^N} \qquad (6.2)$$

---

1. If the quantized signal is passed through an inverse quantization function, the result is not an exact copy of the original signal.

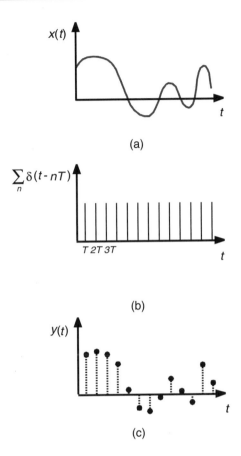

Figure 6.2 Sampling process: (a) input waveform, (b) sampling signal, and (c) sampled output.

where V_{max} is the maximum input level, V_{min} is the minimum input level, and N is the number of bits in each digital code used to describe $y(t)$.

The quantization error is an error sequence $e(i)$ that is dependent on the input signal. It is described by

$$e(i) = x(i) - y(i) \tag{6.3}$$

where $x(i)$ is the analog input and $y(i)$ is the quantization output. If the input is arbitrary (i.e., it crosses many quantization levels), the error sequence can be considered stationary and uncorrelated [2]. As a result, the error is commonly modeled as a random variable with uniform pdf over the range of $-\Delta/2$ to $+\Delta/2$. Additionally, it is assumed to have a psd that is flat from $-f_s/2$ to $f_s/2$,

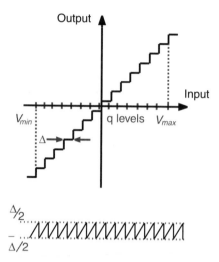

Figure 6.3 Transfer function for an A/D converter.

where f_s is the sampling frequency and is equal to $1/T$, as shown in Figure 6.4.

The variance of the quantization error is given by

$$\sigma_n^2 = \int_{-\Delta_2}^{+\Delta_2} e^2 p(e)de = \frac{\Delta^2}{12} \qquad (6.4)$$

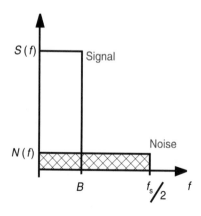

Figure 6.4 The psd of the A/D converter quantization error.

where e is the uniformly distributed error and $p(e)$ is the probability associated with e. The noise power in the bandwidth B is simply

$$v_n^2 = \sigma_n^2 \frac{2B}{f_s} \tag{6.5}$$

The signal power applied to the A/D converter is given by [3]

$$v_s^2 = \alpha^2 \left(\frac{2^N \Delta}{2}\right)^2 \tag{6.6}$$

where α is the loading factor. The loading factor describes the amplitude distribution of the input signal, where

$$\alpha = \frac{V_{rms}}{V_{FS}} \tag{6.7}$$

and V_{FS}, the full-scale voltage, equals $V_{max} - V_{min}$. An N-bit A/D converter accepts an input voltage equal to $2^N \Delta$ without clipping.[2] Note that the A/D converter is fully loaded when the peak-to-peak amplitude of the input signal equals the full-scale range of the converter, and the A/D converter is overloaded when the input amplitude exceeds its full-scale range. Table 6.1 compares the loading factors for some common signals.

SNR generally is used to characterize the performance of an A/D converter. The theoretical limit for the SNR based on an ideal A/D converter is [3]

$$SNR = \frac{v_s^2}{v_n^2} = 3\alpha^2 2^{2N} \left(\frac{f_s}{2B}\right) \tag{6.8}$$

Table 6.1
Loading Factors for Some Common Signals [3]

Signal	Loading Factor (α)
Sine wave	0.707
QPSK modulation	0.5
OQPSK modulation	0.55
Gaussian noise	0.289

---

2. The input signal appears to limit when it exceeds the input range of the A/D converter.

which is often rewritten for convenience in logarithmic terms as

$$SNR = 6.02N + 20\log\alpha^2 + 10\log\left(\frac{f_s}{2B}\right) + 4.77 \qquad (6.9)$$

Improving the SNR of the A/D converter requires greater resolution (N) or a higher sampling rate (f_s). The first effect reduces Δ, the spacing between quantization thresholds, and thus the quantization error. The second effect may not be as obvious; it spreads the quantization noise power to a wider bandwidth and thereby lowers the in-band noise power.

6.1.2 Nonideal Effects

Several effects reduce performance below the theoretical level, including jitter, aliasing, level errors, offset and gain error, circuit noise, and distortion.

The ideal sampling train consists of equally spaced impulse functions. In practical systems, noise disturbs the timing of the pulses; creates uncertainty in the sampling instant, known as jitter; and leads to distortion of the sampled waveform. Jitter is generally modeled as a random variable.

Jitter changes the sampled value of the input signal $x(t)$ by an amount ΔV equal to

$$\Delta V \approx \frac{dx(t)}{dt}\Delta t \qquad (6.10)$$

where Δt is the change or jitter in the sampling instant. That effect is illustrated in Figure 6.5 for a sine wave input signal, that is, $x(t) = A\cos\omega t$. In that situation, the error signal is

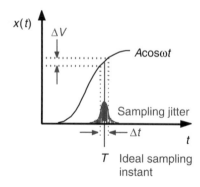

Figure 6.5 Effect of jitter on the sampling process.

$$e(t) = A\omega \sin(\omega t)\Delta t \tag{6.11}$$

If Δt and f are independent, then the error due to sampling jitter is [4]

$$\sigma_e^2 = \frac{1}{2}(\omega A)^2 \sigma_j^2 \tag{6.12}$$

where σ_j is the rms sampling jitter normalized to sampling frequency f_s.

The sampling process folds, or aliases, out-of-band noise and other components to the signal bandwidth. To minimize the folded energy and thus maximize the SNR, a low-pass or antialiasing filter typically precedes the sampling process.

The quantization process is not ideal in practical A/D converters. That is because it is impossible to exactly set the quantization thresholds and to ideally perform the mapping process. Element mismatches and circuit offsets alter the actual quantization levels and create level errors. The variance of these level errors is equal to [5]

$$\sigma_e^2 = \frac{1}{L-1} \sum_{k=1}^{L-1} e_k^2 \tag{6.13}$$

where $L - 1$ is the number of thresholds (equal to $2^N - 1$), and e is the difference between the ideal threshold and the measured threshold.

The ideal thresholds for the level errors analysis are computed from the actual full-scale range and dc offset of the A/D converter. Those parameters scale and shift—but do not distort—the converter's transfer function. As such, the effects of those parameters are correctable by digital means and, therefore, SNR is not degraded.

All these effects are uncorrelated, and, as such, their variances add together to equal the noise plus distortion generated by the A/D converter. The error sum ultimately sets the performance limit of the A/D converter.

6.2 A/D Converter Architectures

A variety of architectures have been developed to combat the nonideal effects that limit resolution and sampling speed in A/D converters. Generally, these architectures fall into one of two categories: Nyquist converters and noise-shaping converters [6]. Nyquist rate converters operate near the Nyquist criterion ($f_s \approx 2B$), while noise-shaping converters use oversampling methods ($f_s \gg 2B$) to improve resolution.

There are actually several different Nyquist A/D converter architectures. Each architecture emphasizes a different feature: fast conversion time, high accuracy, or low power consumption. In practice, Nyquist flash A/D converters translate the I and Q components of the radio signal to digital form and provide multiple samples per chip with 2- to 6-bit resolution.[3] By contrast, noise shaping $\Delta\Sigma$ modulator A/D converters convert audio signals and produce 13- to 16-bit data at an 8-kHz rate [7–9].

Table 6.2 lists the features of some popular A/D converters.

6.2.1 Parallel A/D Converters

The parallel A/D converter, commonly known as the flash A/D converter, offers the fastest conversion times with the lowest latency. It consists of a reference voltage string that connects to a bank of comparators, as shown in Figure 6.6. The flash converter simultaneously compares the analog input signal to each level in the reference string. The sampling and quantizing processes occur at the instant the comparators are "strobed" and produce a thermometer output code (also known as a Gray code [10]) indicating the amplitude of the input signal. The code is then converted to binary format using a simple error check routine and straightforward decoding logic.

An N-bit flash A/D converter requires $2^N - 1$ reference levels and an equal number of comparators. Consequently, each extra bit of resolution doubles the size of the flash converter, the number of critical components, and the input capacitance.

Table 6.2
Comparison of A/D Converter Architectures

Architecture	Benefits	Drawbacks
Parallel	Fast conversion speed, low latency	High power consumption, limited resolution, high device count, large input capacitance
Multistage	Fast conversion, error correction	D/A converter, S/H amplifier, difference circuit
Algorithmic	Excellent accuracy, minimum components	Slow speed, low-droop S/H amplifier, high-accuracy D/A converter
Noise shaping	Excellent accuracy, low-precision analog components	Slow speed, digital filters needed

3. These A/D converters operate above the Nyquist rate to support pilot acquisition by the searcher and bit synchronization by the DLL.

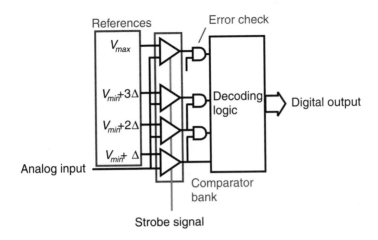

Figure 6.6 Flash A/D converter architecture.

Reference string inaccuracies and comparator offsets created by element mismatches and strobe signal jitter set the performance limit of the flash A/D converter to about 8 bits [4].

6.2.2 Multistage A/D Converters

The multistage A/D converter provides similar high-speed conversion times but increases resolution by using multiple quantizing steps [11]. It uses feedforward (pipelined) or feedback (recursive) structures, as illustrated in Figure 6.7. These A/D converters typically employ two quantizing steps, one for coarse resolution (most significant bits of the digital word) and the other for fine resolution (least significant bits).

The pipelined A/D converter uses separate structures for the coarse and fine quantization steps. The coarse quantizer feeds a D/A converter, which translates the coarse data into an analog signal. The analog signal is then subtracted from the input signal to form the residue signal, which is converted by the fine quantizer.

The feedback A/D converter uses the same quantizer for both the coarse and the fine steps. It alternately selects the analog input or the residue signal for conversion and therefore runs at half the rate of the pipelined structure.

These multistep A/D converters use fewer comparators than flash A/D converters, but they add several crucial analog processing circuits, namely, a sample/hold amplifier, a D/A converter, a difference circuit, and a scaling circuit. Those circuits often set the performance limit of the A/D converter.

The sample/hold amplifier performs the sampling operation using the simple circuit shown in Figure 6.8 [11]. This circuit includes input and output

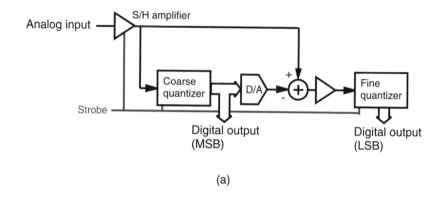

(a)

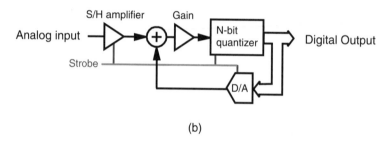

(b)

Figure 6.7 Multistage A/D converters: (a) pipelined structure and (b) feedback structure.

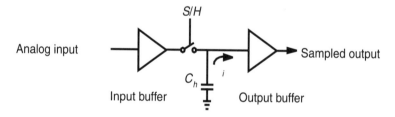

Figure 6.8 Sample/hold amplifier for interfacing to multistage A/D converters.

buffers, a hold capacitor C_h, and a sampling switch. The sample/hold amplifier produces a continuous-time output waveform, not a discrete-time signal, as shown in Figure 6.9. During sampling mode, the input buffer tracks the input signal and charges the hold capacitor through the closed switch. During hold mode, the switch opens and isolates the hold capacitor from the input signal. The output amplifier buffers the sampled value and drives the A/D converter.

Several nonideal effects plague the sample/hold amplifier and are illustrated in Figure 6.10. In practice, the sampling switch possesses parasitic capacitance. Charge stored by that capacitance is injected onto the hold capacitor

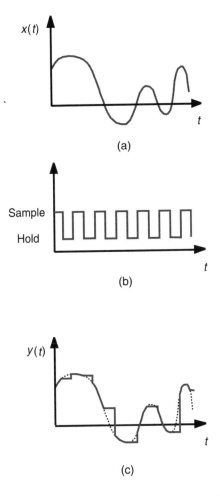

Figure 6.9 Operation of the sample/hold amplifier: (a) input signal, (b) sample/hold signal, and (c) continuous-time output waveform.

when the sample/hold amplifier is switched from sampling mode to hold mode. The charge causes the voltage stored on the capacitor to jump (an effect known as "hold jump") and creates the pedestal in the output waveform. The parasitic capacitance of the switch also prevents complete isolation of the hold capacitor from the input signal. Consequently, a small fraction of the input signal appears at the output during hold mode, an effect referred to as "hold-mode feedthrough."

Ideally, for multistage A/D converters, the sample/hold amplifier keeps the output voltage constant during hold mode and allows the A/D converter to perform multiple quantizing steps. Any change in the input signal between

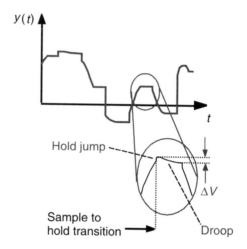

Figure 6.10 Some nonideal effects in the sample/hold amplifier.

sampling operations introduces an error. In practice, a finite input current i flows from the hold capacitor to the output buffer and lowers the voltage stored on the capacitor by an amount equal to

$$\Delta V = \frac{i}{C_h} \Delta t \qquad (6.14)$$

where ΔV is the "droop" in stored voltage and Δt is the elapsed time between the coarse and fine quantizing steps. Note that larger values of C_h are better at holding the sampled signal but are more difficult to drive during sampling mode.

6.2.3 Algorithmic A/D Converters

The algorithmic A/D converter uses a single 1-bit quantizer in a recirculating mode. This approach requires very few components and can produce extremely accurate results, but it operates slowly. In fact, its conversion rate is inversely proportional to resolution (N).

   The successive approximation A/D converter is an example of an algorithmic converter and is shown in Figure 6.11(a). It converts the analog input sample to a digital word one bit at a time. After each quantizing operation, the residue is formed, scaled, and reapplied to the quantizer (shown as a simple comparator). Because any error is magnified through the recirculating process, careful design is critical. The successive approximation A/D converter trades higher resolution for slower conversion speed.

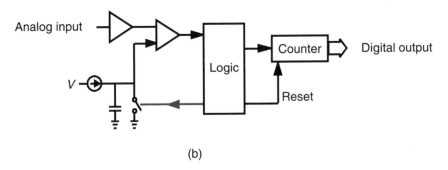

(a)

(b)

Figure 6.11 Algorithmic A/D converters: (a) successive approximation architecture and (b) integrating architecture.

Another example of an algorithmic converter is the integrating A/D converter shown in Figure 6.11(b). It operates as follows. The quantization process starts with the counter cleared and the switch opened. That allows the capacitor to charge, creating a linear voltage ramp that eventually crosses the input signal. When that happens, the logic detects the change in the comparator's output and stops the counter.

It takes M clock cycles for the capacitor to charge to the full-scale input of the A/D converter. Therefore, the digital output code of the integrating A/D converter is simply

$$y(n) = \frac{v_a(nT)}{V_{FS}} M \qquad (6.15)$$

where $v_a(nT)$ is the analog input voltage, sampled at nT.

6.2.4 Noise-Shaping A/D Converters

In a conventional A/D converter, the performance depends on the number and uniformity of the quantization levels and the oversampling ratio (f_s/B). As N increases, the difference between levels (Δ) shrinks and analog precision becomes more critical. As f_s/B increases beyond the Nyquist value, the spectral density of the quantization error decreases relatively slowly.

A more efficient oversampling quantizer is the $\Delta\Sigma$ modulator [12–15], shown in Figure 6.12(a). This is a first-order modulator with an analog filter $H(s)$ and a single-bit data converter. The oversampled A/D converter shapes the spectrum of the quantization error, thereby significantly improving the in-band SNR.

The signal transfer function for the circuit is a low-pass filter response that extends to the band edge (B) of the analog signal, where

$$S(s) = \frac{H(s)}{1 + H(s)} \tag{6.16}$$

and the filter $H(s)$ serves as an integrator. The quantization noise transfer function is a high-pass filter response that extends to $+f_s/2$, where

$$N(s) = \frac{1}{1 + H(s)} \tag{6.17}$$

It pushes quantization noise outside the signal bandwidth B toward the Nyquist frequency $f_s/2$ and demonstrates the noise-shaping property of the $\Delta\Sigma$ modulator shown in Figure 6.12(b).

$\Delta\Sigma$ modulators operate at very high sampling rates, which leads to the alternative label *oversampling converters*. In the $\Delta\Sigma$ modulator, the integrator $H(s)$ accumulates the difference, or error signal, between the input signal and the quantized value. The error signal is driven toward zero by the feedback loop, producing a bit stream output with a duty cycle equal to the amplitude of the input. In fact, it is that feedback that is the key to the improved efficiency of the $\Delta\Sigma$ modulator. A digital filter removes the quantization noise from the bit stream in the frequency band of B to $f_s/2$ to provide a wider dynamic range output. As a result, the $\Delta\Sigma$ modulator provides superior performance with low-cost, imprecise analog components.

The first-order $\Delta\Sigma$ modulator is mathematically analyzed with the discrete-time model of Figure 6.12(c). In that model, an accumulator replaces the integrator. The output of the accumulator is

$$w(i) = x(i - 1) - e(i - 1) \tag{6.18}$$

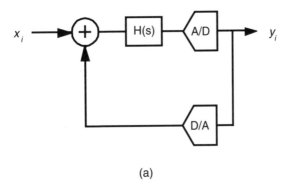

(a)

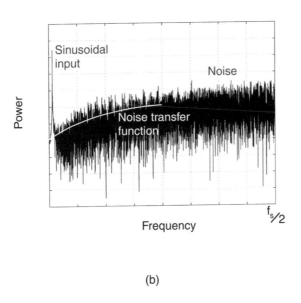

(b)

Figure 6.12 First-order oversampled A/D converter or $\Delta\Sigma$ modulator: (a) architecture, (b) spectrum of the quantization noise, and (c) mathematical model.

which yields the quantized signal

$$y(i) = w(i) + e(i) = x(i - 1) + [e(i) - e(i - 1)] \qquad (6.19)$$

In (6.19) and (6.20), x is the continuous time analog signal at the sampling instant, w is the multibit output of the accumulator, e is the error associated with the single-bit quantizer, and y is the single-bit output.

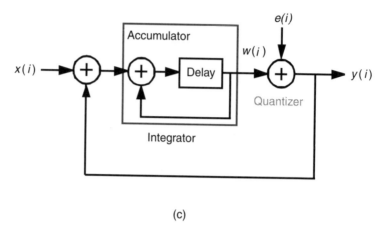

(c)

Figure 6.12 (continued).

This circuit treats the quantization error and the input signal differently. The output is the first difference of the quantization error, while the input signal is unchanged, except for a delay. To calculate the effective resolution of the $\Delta\Sigma$ modulator, the input signal is assumed to be significantly busy so that the error can be treated as white noise and uncorrelated with the input signal [16, 17]. The modulation noise is defined as $n(i) = e(i) - e(i-1)$ and can be expressed by the z-transform

$$N(z) = E(z)[1 - z^{-1}] \tag{6.20}$$

If the transfer function of the integrator, $H(z)$, is defined by $[1 - z^{-1}]^{-1}$ and $z = e^{jwT}$, then $H(f) = 2\sin(\pi fT)$. This is used to rewrite (6.20) in the form of the rms output quantization noise voltage spectral density, that is,

$$N(f) = 2E(f)\sin(\pi fT) \tag{6.21}$$

where $E(f)$ is the spectral density of the single-bit standalone quantizer.
The equivalent noise power in the band of interest is

$$n^2 = \int_0^B \sigma_n^2 2T[2\sin(\pi fT)]^2 df \tag{6.22}$$

which is evaluated using a series expansion for the sin term. That produces the result

$$n^2 = \frac{8}{3}\pi^2 \sigma_n^2 (TB)^3 \qquad (6.23)$$

and shows that the noise power falls off 9 dB for each octave of oversampling improvement. (In a conventional A/D converter, that factor is 3 dB.)

Another common $\Delta\Sigma$ modulator is the second-order architecture [18, 19] shown in Figure 6.13(a). Here, the quantized signal is

$$y(i) = x(i-1) + [e(i) - 2e(i-1) - e(i-2)] \qquad (6.24)$$

and the modulation noise is $n(i) = e(i) - 2e(i-1) + e(i-2)$. This has a z-transform given by

$$N(z) = E(z)[1 - z^{-1}]^2 \qquad (6.25)$$

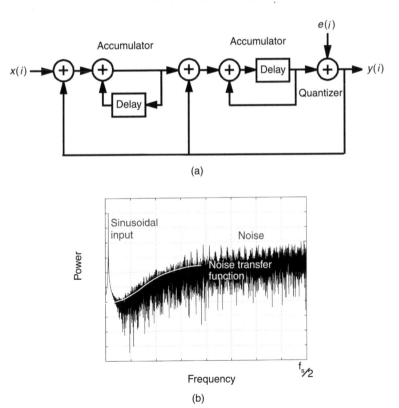

(a)

(b)

Figure 6.13 Second-order $\Delta\Sigma$ modulator: (a) discrete-time model and (b) noise-shaping property.

The spectral density of the rms noise voltage for this architecture is

$$N(f) = E(f)[2\sin(\pi ft)]^2 \qquad (6.26)$$

and is shown in Figure 6.13(b). As before, the in-band noise power is determined by integration over the band of interest

$$n^2 = \frac{32}{5}\pi^2\sigma_n^2(TB)^5 \qquad (6.27)$$

The second-order $\Delta\Sigma$ modulator provides 15-dB SNR improvement for each doubling of the oversampling ratio.

In a second-order modulator, the input signal and the contents of the accumulators are all integer multiples of the quantizer step. When the input is a dc value, limit cycles are generated. Those limit cycles are perceived as annoying tones. To prevent them, a dither signal (single-bit PN sequence) is often added to the input signal [20].

To take advantage of the noise-shaping benefit provided by the $\Delta\Sigma$ modulator A/D converter, the high-speed digital output from the quantizer must be filtered and decimated. The filtering occurs before decimation and removes excessive noise power from the frequency range B to $f_s/2$; otherwise, the noise aliases to the signal band. It also transforms the high-speed single-bit data stream into a low-speed, high-resolution data stream. In practice, a sinc^k decimation filter is typically used [21], where the order of the filter k is equal to $l + 1$ and l is the order of the $\Delta\Sigma$ modulator. Mathematically, this sinc^k filter is defined as

$$D(f) = \left[\frac{\text{sinc}(\pi fNT')}{\text{sinc}(\pi fT')}\right]^k \qquad (6.28)$$

where $\text{sinc}(x) = \sin(x)/x$. Note that the sinc^k decimation filter slightly raises the in-band noise power.

The noise psd's for the first-order $\Delta\Sigma$ modulator with a sinc^2 filter and for the second-order $\Delta\Sigma$ modulator with a sinc^3 filter are plotted in Figure 6.14.

A $\Delta\Sigma$ modulator A/D converter has a lower overload level than a conventional A/D converter. The reason for the difference is as follows. In the first-order $\Delta\Sigma$ modulator shown in Figure 6.12(a), the single-bit D/A converter generates a two-level feedback signal that toggles at a high rate with an average value equal to the amplitude of the input signal. As such, the two output

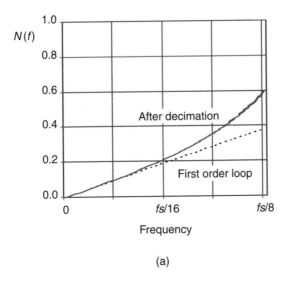

(a)

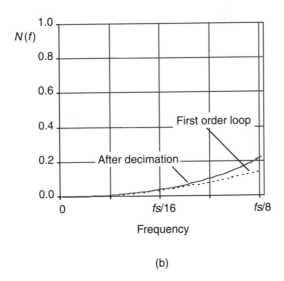

(b)

Figure 6.14 Sinc filter effect (a) first-order $\Delta\Sigma$ modulator decimated by sinc^2 filter and (b) second-order $\Delta\Sigma$ modulator decimated by sinc^3 filter.

levels from the D/A converter set lower and upper limits on the input signal. Furthermore, the output of the summer, which combines the input signal and the feedback signal, can approach twice those minimum and maximum levels. Thus, to prevent overloading of the $\Delta\Sigma$ modulator A/D converter, the amplitude of the analog signal is restricted to one-half the full-scale range [21].

6.3 D/A Conversion

The D/A converter transforms digital data to its analog form by reconstructing and filtering the sampled data. Its performance is fundamentally limited by the resolution and the sampling rate. The D/A conversion process is governed by many of the same principles as the A/D conversion process.

6.3.1 Ideal Process

The ideal D/A converter provides an analog output equal to the digital data applied at its input using the simple structure shown in Figure 6.15. The data generally is latched to trigger the transition between samples in much the same way as the A/D sampling process occurs at a single instant. The output of the D/A converter typically is passed through a low-pass filter to smooth the reconstructed waveform.

Mathematically, the D/A converter multiplies the impulse sampling function with the analog value of the digital input and holds the result to produce the output waveform shown in Figure 6.16. The analog value of the digital input is given by

$$v_a(x) = (x_0 + x_1 2 + x_2 2^2 \ldots x_{N-1} 2^{N-1})\Delta \tag{6.29}$$

where x_i are the individual binary-weighted bits of an N-bit digital word. This expression shows the full-scale range, V_{FS}, of the D/A converter is simply

$$V_{FS} = 2^N \Delta \tag{6.30}$$

Note that the resolution—and thus the quantization error—of an N-bit D/A converter and an N-bit A/D converter are identical.

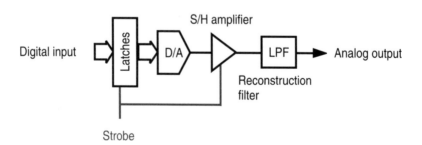

Figure 6.15 Simple D/A converter architecture.

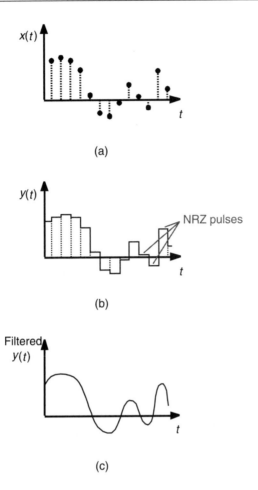

Figure 6.16 D/A converter waveforms: (a) impulse sampled waveform, (b) S/H amplifier output, and (c) filtered output signal.

6.3.2 Nonideal Effects

The D/A converter is plagued by circuit nonidealities that introduce distortion and add frequency components. The nonidealities are similar in origin and effect to those associated with the A/D converter.

The output signal produced by the D/A converter is actually a series of nonreturn-to-zero (NRZ) pulses. In practice, those pulses are not impulse signals[4] but have a nonzero width equal to the conversion rate T. As such, the D/A conversion process can be described in the frequency domain by

---

4. Ideally, the NRZ pulse is an impulse signal.

$$Y(f) = X(f) \cdot \frac{\sin(\pi f T)}{\pi f T} \tag{6.31}$$

where the second term[5] represents the Fourier transform of the NRZ pulse. Equation (6.31) shows that the width of the NRZ pulse affects the frequency spectrum of the D/A converter's output signal. If the D/A converter is operated at or near the Nyquist rate ($B_{x(t)} \approx 1/2T$), the spectrum of $x(t)$ near the edge of the band becomes distorted, as shown in Figure 6.17(a). To prevent that, the D/A converter typically is operated well above the Nyquist rate, as shown in Figure 6.17(b).

Each sampling point switches the analog output to a new value, which typically generates glitches in the output waveform, as shown in Figure 6.18. The glitches are caused by nonideal circuit switches, require a finite time to settle, and limit the conversion speed of the D/A converter.

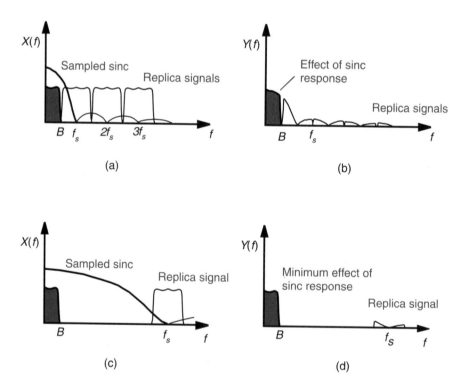

Figure 6.17 Effect of nonimpulse sampling: (a) at Nyquist rate and (b) required increase in sampling frequency to minimize effect of nonzero pulse width.

5. The second term is commonly known as the sample sinc function.

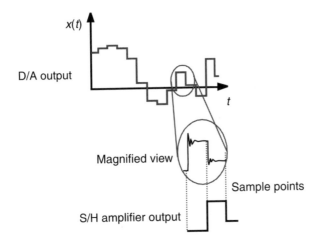

Figure 6.18 D/A converter output glitches.

To improve the fidelity of the output waveform, a sample/hold amplifier follows the D/A converter. It allows the analog output signal to settle and captures the signal before the D/A converter switches to a new value.

The static or low-frequency accuracy of the analog waveform is measured using two figures of merit: differential nonlinearity (DNL), illustrated in Figure 6.19(a), and integral nonlinearity (INL), shown in Figure 6.19(b). DNL measures the difference between analog voltages generated by adjacent digital codes and compares that difference against the ideal step size, that is,

$$\text{DNL} = \frac{y(x_k) - y(x_{k+1})}{\Delta} - 1 \tag{6.32}$$

and is expressed as a fraction of an lsb. Note that if the converter's response decreases for any digital step, the transfer function becomes nonmonotonic and the resolution degrades significantly.

INL measures the overall linearity of the transfer function. It uses a straight-line fit between the zero- and full-scale analog outputs to remove gain and offset effects. It then compares the expected value using that straight-line fit against the actual analog values as follows

$$\text{INL} = \frac{y(x_k) - g(x_k)}{\Delta} \tag{6.33}$$

where the function $g(x)$ describes the straight-line fit

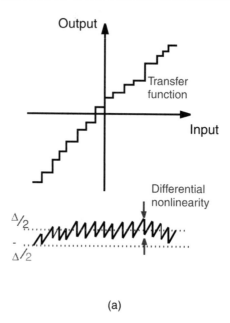

(a)

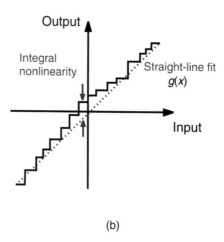

(b)

Figure 6.19 Static nonlinearities in D/A converters: (a) differential nonlinearity and (b) integral nonlinearity.

$$g(x) = \frac{V_{FS}}{2^N}x + V_{min} \qquad (6.34)$$

These nonlinear effects raise the in-band noise power and introduce distortion.

6.4 D/A Converter Architectures

There are only a few architecture options for D/A converters, and they fall into one of two categories: scaling and noise shaping. Scaling architectures provide flexibility and allow fast conversion times and/or high precision, while noise-shaping structures primarily target high-precision applications.

Both types of D/A converters are used to form the radio signals for transmission. In practice, the radio signals are processed as I and Q components with two separate D/A converters. Each converter provides 8-bit analog waveforms with a bandwidth of 615 kHz.[6] By contrast, the audio signal[7] is reproduced by a noise-shaping D/A converter with 10- to 14-bit resolution [22, 23].

6.4.1 Scaling D/A Converter Concepts

Scaling architectures use binary-weighted quantities of current, voltage, or charge to generate an output waveform [24]. For the current-scaling D/A converter shown in Figure 6.20, the digital code selects the current sources that connect to the output.

Very accurate current sources are needed in this scheme and are realized with precise analog components. The binary weights are implemented by device

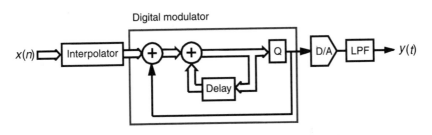

Figure 6.20 Simplified schematic of current-scaling D/A converter.

---

6. These signals are combined in the I/Q modulator to form the wide spread-spectrum waveform.
7. Audio signal bandwidth is almost always less than 4 kHz.

scaling or by multiple equal-weight devices, a technique known as segmenting [24]. The segmented approach achieves better matching and accuracy, but it is impractical for more than 6 bits. In practice, high-end D/A converters use a combination of segmented (for lower lsb's) and binary-weighted scaling.

6.4.2 Oversampled D/A Converters

The $\Delta\Sigma$ modulator concept is also useful for D/A conversion [12]. It operates under similar principles as the $\Delta\Sigma$ modulator A/D converter. An interpolator elevates the sampling frequency of the digital input data to match the oversampling requirement. The summer compares the interpolated input data to the single-bit feedback signal and produces an error signal. The error signal is accumulated (integrated), and the result is fed forward to the quantizer. The quantizer function reduces the multibit data to a single bit, which is routed to the single-bit D/A converter and is also fed back to the summer.

The D/A converter transforms the high-rate single-bit stream to an analog signal with noise-shaped properties. An analog low-pass filter removes the out-of-band quantization noise and smoothes the signal. The result is an analog output with wide dynamic range and extremely low noise floor. The $\Delta\Sigma$ modulator D/A converter provides a high-resolution analog output with low-cost, analog components, and sophisticated digital signal processing.

This D/A converter relies on a digital $\Delta\Sigma$ modulator but adheres to the same principles as the $\Delta\Sigma$ modulator A/D converter.

References

[1]   Taub, H., and D. L. Schilling, *Principles of Communication Systems*, New York: McGraw-Hill, 1986.

[2]   Bennett, W. R., "Spectra of Quantized Signals," *Bell Systems Tech. J.*, Vol. 27, July 1948, pp. 446–472.

[3]   Martin, D. R., and D. J. Secor, *High Speed Analog-to-Digital Converters in Communication Systems*, Nov. 1981.

[4]   Frerking, M. E., *Digital Signal Processing in Communication Systems*, Boston: Kluwer Academic Publishers, 1994.

[5]   Oyama, B., "Guidelines for A/D and D/A Converters Error Budgets," July 1979.

[6]   Khoury, J., and H. Tao, "Data Converters for Communication Systems," *IEEE Communications Magazine*, Oct. 1998, pp. 113–117.

[7]   Matsumoto, K., et al., "An 18b Oversampling A/D Converter for Digital Audio," *ISSCC Digest of Technical Papers*, Feb. 1988, pp. 202–203.

[8]   Norsworthy, S. R., I. G. Post, and H. S. Fetterman, "A 14-bit 80-kHz Sigma-Delta A/D Converter: Modeling, Design, and Performance Evaluation," *IEEE J. of Solid-State Circuits*, Vol. SC-24, Apr. 1989, pp. 256–266.

[9]   Welland, D. R., et al., "A Stereo 16-Bit Delta-Sigma A/D Converter for Digital Audio," *J. of Audio Engineering Society*, Vol. 37, June 1989, pp. 476–486.

[10]   Proakis, J. G., *Digital Communications*, New York: McGraw Hill, 1995.

[11]   Razavi, B., *Principles of Data Conversion System Design*, New York: IEEE Press, 1995.

[12]   Candy, J. C., and G. C. Temes, "Oversampling Methods for A/D and D/A Conversion," *Oversampling Delta-Sigma Data Converters*, New York: IEEE Press, 1992.

[13]   Agrawal, B. P., and K. Shenoi, "Design Methodology for $\Sigma\Delta M$," *IEEE Trans. on Communications*, Vol. COM-31, No. 3, Mar. 1983, pp. 360–369.

[14]   Tewksbury, S. K., and R. W. Hallock, "Oversampled, Linear Predictive and Noise-Shaping Coders of Order N > 1," *IEEE Trans. on Circuits and Systems*, Vol. CAS-25, No. 7, July 1978, pp. 436–447.

[15]   Candy, J. C., "A Use of Limit Cycle Oscillations to Obtain Robust Analog-to-Digital Converters," *IEEE Trans. on Communications*, Vol. COM-22, No. 3, Mar. 1974, pp. 298–305.

[16]   Gray, R. M., "Quantization Noise Spectra," *IEEE Trans. on Information Theory*, Vol. IT-36, Nov. 1990, pp. 1220–1244.

[17]   Candy, J. C., and O. J. Benjamin, "The Structure of Quantization Noise From Sigma-Delta Modulation," *IEEE Trans. on Communications*, Vol. COM-29, Sept. 1981, pp. 1316–1323.

[18]   Candy, J. C., "A Use of Double Integration in Sigma Delta Modulation," *IEEE Trans. on Communications*, Vol. COM-33, Mar. 1985, pp. 249–258.

[19]   Boser, B. E., and B. A. Wooley, "The Design of Sigma-Delta Modulation Analog-to-Digital Converters," *IEEE J. of Solid-State Circuits*, Vol. SC-23, Dec. 1988, pp. 1298–1308.

[20]   Chou, W., and R. M. Gray, "Dithering and Its Effects on Sigma-Delta and Multi-Stage Sigma-Delta Modulation," *IEEE Proc. for ISCAS '90*, May 1990, pp. 368–371.

[21]   Candy, J. C., "Decimation for Sigma Delta Modulation," *IEEE Trans. on Communications*, Vol. 34, Jan. 1986, pp. 72–76.

[22]   Carley, L. R., and J. Kenney, "A 16-Bit 4th Order Noise-Shaping D/A Converter," *IEEE Proc. of CICC*, 1988, pp. 21.7.1–21.7.4.

[23]   Naus, P. J. A., et al., "A CMOS Stereo 16-Bit D/A Converter for Digital Audio," *IEEE J. of Solid-State Circuits*, Vol. SC-22, June 1987, pp. 390–395.

[24]   Grebene, A. B., *Bipolar and MOS Analog Integrated Circuit Design*, New York: Wiley, 1984.

7

RF System Fundamentals

The RF transceiver provides the wireless link for untethered communications. It establishes forward-link and reverse-link communication channels using radio spectrum in designated bands between 800 MHz and 2,000 MHz. Those radio frequencies enable efficient wireless communications with tolerable path losses, reasonable transmit power levels, and practical antenna dimensions.

Figure 7.1 is a simplified view of a CDMA RF transceiver. It consists of a transmitter, a receiver, and a frequency synthesizer. The transmitter shifts baseband signals to the assigned radio frequency using a two-step heterodyne architecture. The receiver selects the designated channel and translates its carrier frequency to baseband using a similar architecture. The frequency synthesizer

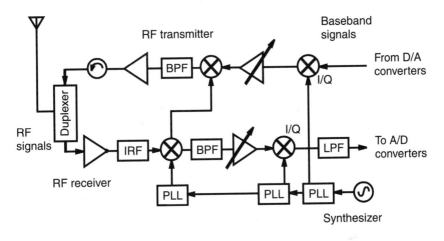

Figure 7.1 Block diagram of a simplified RF transceiver.

149

provides the reference signals needed for frequency translation in the transmitter and the receiver.

This chapter reviews the RF concepts of duplex operation, frequency translation, phase modulation, dynamic range, and frequency synthesis. It describes the operation of the transmitter, including power control, spurious response, and other performance requirements. It concludes with a discussion of critical issues in the receiver, such as sensitivity, desensitization, and frame error rate.

7.1 RF Engineering Concepts

The RF system plays a key role in the battle against deleterious radio propagation effects. The transmitter conditions the baseband signals for efficient wireless communications, while the receiver selects the desired message signal from various interfering signals. To perform those operations requires a duplex arrangement with frequency translation and phase modulation functions. It also requires a wide operating range, bounded by background noise at low signal levels and nonlinear effects at high signal levels. The following sections outline the operation of those functions and their limitations.

7.1.1 Duplex Operation

One-way communication, from a single source to a single destination, is known as a simplex arrangement. Two-way communication, involving forward and reverse communication links, is referred to as a duplex system. In time division duplex (TDD) systems, like GSM, the transmitter and the receiver operate at different times [1]. In frequency division duplex (FDD) networks, such as CDMA IS95, the transmitter and the receiver function simultaneously using separate radio channels [1].[1]

Typically, FDD systems require greater than 120 dB isolation from the transmitter to the receiver. The reason is that noise generated by the transmitter appears at the antenna, elevates the receiver's noise floor, and thereby lessens the receiver's ability to detect small signals.

A duplex filter at the antenna isolates the transmitter from the receiver. It combines a transmit-band filter and a receiver-band filter, as shown in Figure 7.2. This structure generally provides the necessary isolation between the two paths but slightly attenuates the in-band signals, lowering the power of the

---

1. It is not uncommon for networks that support transmission and reception at the same time to be referred to as full duplex.

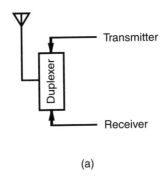

(a)

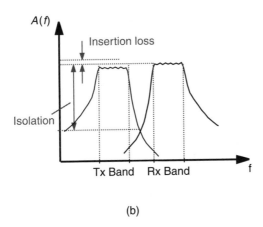

(b)

Figure 7.2 Duplex filter: (a) schematic diagram and (b) frequency response.

transmitted signal and raising the noise level of the received signal. The duplex filter also occupies a large amount of physical board space, typically more than any other component.

7.1.2 Frequency Translation

A key process of the RF transceiver is frequency translation. To communicate efficiently via wireless channels with acceptably sized antennas, microwave signals are used. Microwave signals are formed by shifting the carrier of the modulated signal from baseband to radio frequencies in the transmitter. A similar, but reverse, translation process is employed to shift radio signals to baseband in the receiver.

Mixer circuits are used for frequency translation. These circuits are functionally equivalent to analog multipliers that linearly multiply two input signals together to produce an output signal described by

$$s(t) = A\cos(2\pi f_1 t) \times \cos(2\pi f_2 t) \qquad (7.1)$$

where f_1 is the input signal to be shifted and f_2 is the local oscillator (LO) signal. (The LO signal is generated specifically by the frequency synthesizer for frequency translation.) Equation (7.1) can be rewritten as

$$s(t) = \frac{A}{2}[\cos 2\pi(f_1 - f_2)t + \cos 2\pi(f_1 + f_2)t] \qquad (7.2)$$

which shows output components at the sum and the difference of the input frequencies. It also shows that in an ideal mixer the amplitude of the output signal is proportional to the amplitude of the input signal.

7.1.3 Phase Modulation

An important step in the channel coding process is modulation, which superimposes the message signal onto the carrier waveform. The modulation process changes the amplitude and/or the phase angle of the carrier signal [2–4]. Different modulation schemes offer various advantages in terms of bandwidth efficiency and simplicity. The CDMA IS95 system uses QPSK modulation, as illustrated in Figure 7.3.

The phase-modulated signal is described by

$$s(t) = A_c\cos[\omega_c t + \theta(t)] \qquad (7.3)$$

where A_c is the amplitude of the signal, ω_c is the carrier radian frequency, $\theta(t)$ is the time-varying phase function equal to $\kappa m(t)$, κ is the phase sensitivity of modulation in rads per volt, and $m(t)$ is the binary message signal. Equation (7.3) can be expanded to

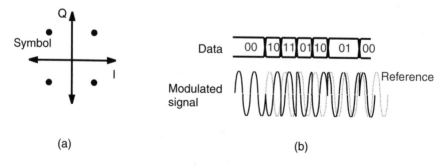

(a)   (b)

Figure 7.3 QPSK: (a) constellation diagram and (b) waveform.

$$s(t) = \frac{A_c}{\sqrt{2}}\cos\omega_c t \cos\kappa m(t) - \frac{A_c}{\sqrt{2}}\sin\omega_c t \sin\kappa m(t) \qquad (7.4)$$

where the factors with $\omega_c t$ correspond to the pilot or carrier waveform and the factors with $\kappa m(t)$ indicate the modulated data.

For QPSK signaling, each message symbol $m(t)$ represents two bits of data, with orthogonal values of $1+j$, $-1+j$, $-1-j$, and $1-j$. The complex envelope of the QPSK-modulated signal is

$$g(t) = A_c e^{j\theta(t)} \qquad (7.5)$$

which conveniently can be rewritten as

$$g(t) = x(t) + jy(t) \qquad (7.6)$$

where $x(t) = A_c/\sqrt{2}\cos\theta(t)$ and $y(t) = jA_c/\sqrt{2}\sin\theta(t)$. Note that the psd of the QPSK-modulated signal is simply the spectral shape of the message signal $m(t)$. If the message signal consists of rectangular pulses, then

$$S(f) = A_c^2 T_b \left(\frac{\sin \pi f T_b}{\pi f T_b} \right) \qquad (7.7)$$

where T_b is the symbol period of the message.

The QPSK modulator is based on (7.4) and is formed using the structure of Figure 7.4. It relies on orthogonal signals to drive two multipliers, which requires decomposing the message signal into orthogonal components, $A_c/\sqrt{2}\cos\kappa m(t)$ and $A_c/\sqrt{2}\sin\kappa m(t)$. The decomposition proves easy and makes phase modulation by this technique both efficient and convenient.

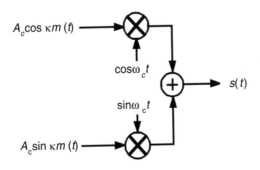

Figure 7.4 Block diagram of QPSK modulator.

The phase-modulated signal is recovered by using the complimentary structure known as an I/Q demodulator, shown in Figure 7.5. It takes the received signal $r(t)$ and multiplies it by orthogonal carrier signals, $\cos\omega_c t$ and $\sin\omega_c t$, whose frequency and phase have been precisely aligned to the incoming signal. The received signal is described by

$$r(t) = A_c(m_I \cos\omega_c t + m_Q \sin\omega_c t) \qquad (7.8)$$

where m_I and m_Q are the I and Q components of the message signal, respectively. The outputs of the mixers are

$$r_I(t) = A_c m_I \cos\omega_c t \cos\omega_c t + A_c m_Q \sin\omega_c t \cos\omega_c t \qquad (7.9a)$$
$$r_Q(t) = A_c m_I \cos\omega_c t \sin\omega_c t + A_c m_Q \sin\omega_c t \sin\omega_c t \qquad (7.9b)$$

which expands by trigonometric identities to yield

$$r_I(t) = \frac{m_I}{2}[\cos(0) + \cos(2\omega_c t)] + \frac{m_Q}{2}[\sin(2\omega_c t) + \sin(0)] \quad (7.10a)$$

$$r_Q(t) = \frac{m_I}{2}[\sin(2\omega_c t) - \sin(0)] + \frac{m_Q}{2}[\cos(0) - \cos(2\omega_c t)] \quad (7.10b)$$

Low-pass filters remove the double frequency terms, $2\omega_c t$, and leave the original message signals $m_I(t)$ and $m_Q(t)$.

7.1.4 Noise

All electronic circuits are plagued by noise, which arises from the thermal agitation of electrons as well as the discrete nature of current flow [5]. As a

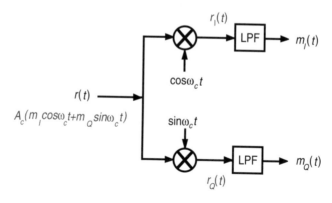

Figure 7.5 Block diagram of QPSK demodulator.

result, current flow shows tiny fluctuations that are essentially random and noiselike. This is important because noise raises the minimum allowed signal level and thereby affects system performance.

Noise is typically characterized in an RF circuit by the parameter *Noise Factor* (F) and is defined by

$$F = \frac{\left(\dfrac{S}{N}\right)_{in}}{\left(\dfrac{S}{N}\right)_{out}} \tag{7.11}$$

where $(S/N)_{in}$ is the ratio of signal power to noise power available at the input to the circuit and $(S/N)_{out}$ is that ratio at the output of the circuit, as shown in Figure 7.6(a). The quantity F is often expressed as the parameter *Noise Figure* (NF), which is related by the expression $NF = 10 \log F$. Because the

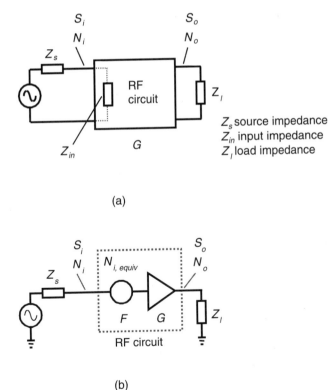

(a)

(b)

Figure 7.6 Noise factor: (a) definition and (b) model.

gain of the circuit $G = S_o/S_i$, the noise factor can be rewritten as $F = N_o/GN_i$. The input noise power N_i is related to the available power from a source at the thermal noise floor, $4kTB$, where k is Boltzman's constant, T, is the absolute temperature of the reference,[2] and B is the bandwidth of the measurement. The noise power delivered to the RF circuit, when its input is conjugately matched to the source, is the available power, kTB.

Another way to look at noise factor is to model the circuit as an ideal noiseless amplifier preceded by an equivalent noise generator, as shown in Figure 7.6(b). Then

$$N_{i,equiv} = \frac{N_o}{G} = kTFB \qquad (7.12)$$

But the output noise is just the sum of the amplified input noise ($N_i = kTB$) and the circuit noise (N_x), as shown by the equation

$$N_o = GN_i + N_x \qquad (7.13)$$

Therefore, the circuit noise is simply the excess noise $N_x = kTB(F-1)$.

In a cascaded system, like the one shown in Figure 7.7, the noise power at the output of the first stage is given by

$$N_{o1} = G_1 kTF_1 B \qquad (7.14)$$

and the input-referred circuit noise of the second stage is

$$N_{x2} = kT(F_2 - 1) \qquad (7.15)$$

Combining those two noise powers and multiplying by the gain of the second stage yields the noise power at the output of the second stage:

$$N_{o2} = G_1 G_2 kTF_1 + G_2 kT(F_2 - 1) \qquad (7.16)$$

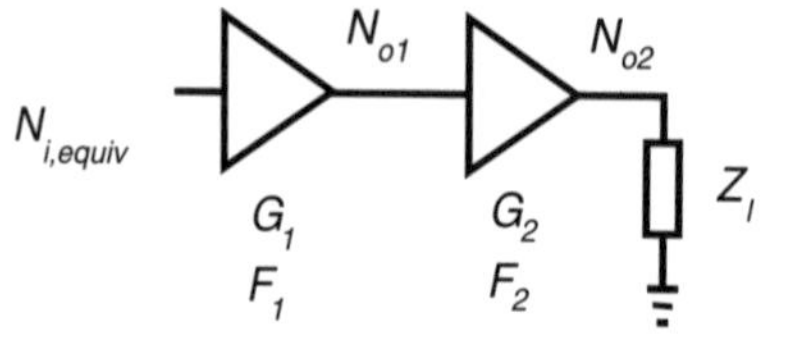

Figure 7.7 Calculation of cascaded noise factor.

2. The typical value of T for noise figure measurements is 290K [6].

The noise factor of the cascaded system F_T is found by dividing this expression by the factors kTB and the cascaded power gain, $G_1 G_2$

$$F_T = F_1 + \frac{F_2 - 1}{G_1} \tag{7.17}$$

From that analysis, it is straightforward to expand (7.17) to Friis's well-known expression for noise factor in an n-stage cascaded system [7]:

$$F_T = F_1 + \frac{F_2 - 1}{G_1} + \ldots + \frac{F_n - 1}{\displaystyle\prod_{i=1}^{n-1} G_i} \tag{7.18}$$

Note that the effect on the overall system noise factor of later stages is reduced by the gain of earlier stages. That means the first stage (or first few stages) essentially determines the cascaded noise factor if its gain is sufficiently high. The cascaded noise factor shows the excess noise produced by the system and thereby sets the minimum detectable signal level.

7.1.5 Distortion

The operating range, or dynamic range, of the system is limited at high signal levels by nonlinear circuit effects, including gain compression, harmonic distortion, and intermodulation distortion. Those effects muddle the message signal and degrade its quality. They also introduce a subtle effect, in which signals at one frequency can influence signals at another frequency.

The output voltage of an RF circuit can be modeled by a power series expansion of its nonlinear gain using the form

$$v_o = a_1 v_i + a_2 v_i^2 + a_3 v_i^3 + \ldots \tag{7.19}$$

where v_o is the output signal, v_i is the input signal, and coefficients a_1, a_2, and a_3 are frequency dependent. If an input signal $V\cos\omega t$ is applied to the RF circuit, it will produce an output signal equal to

$$v_o(t) = a_1 V \cos\omega t + a_2 V^2 \cos^2 \omega t + a_3 V^3 \cos^3 \omega t + \ldots \tag{7.20}$$

which expands to

$$v_o(t) = a_1 V \cos\omega t + \frac{a_2}{2} V^2 (1 + \cos 2\omega t) + \frac{a_3}{4} V^3 (3\cos\omega t + \cos 3\omega t) + \ldots \tag{7.21}$$

which shows the mechanisms for gain compression and harmonic distortion.

The RF circuit amplifies the input signal $V\cos\omega t$ and generates an output signal at the fundamental frequency ω with an amplitude equal to

$$V_1 = a_1 V + \frac{3}{4} a_3 V^3 \tag{7.22}$$

where coefficient a_1 represents the small-signal gain, and coefficient a_3 causes gain compression.[3] The gain compression α, expressed as a fraction of the linear gain, is described by

$$\alpha = \frac{V_1}{a_1 V} = 1 + \frac{3a_3}{4a_1} V^2 \tag{7.23}$$

and is typically reported at the -1 dB point ($\alpha = 0.891$). This 1-dB compression point occurs when

$$V = 0.381 \sqrt{\left|\frac{a_1}{a_3}\right|} \tag{7.24}$$

The RF circuit also produces output signals at frequencies that are integer multiples of the fundamental frequency, which is known as harmonic distortion, that is,

$$V_2 = \frac{a_2}{2} V^2 \cos 2\omega t \qquad V_3 = \frac{a_3}{4} V^3 \cos 3\omega t \tag{7.25}$$

The relative strengths of those signals are

$$\frac{V_2}{V_1} = \frac{a_2}{2a_1} V \qquad \frac{V_3}{V_1} = \frac{a_3}{4a_1} V^2 \tag{7.26}$$

and are dependent on the amplitude of the input signal.

If the input consists of two signals at different frequencies, $V_1 \cos \omega_1 t + V_2 \cos \omega_2 t$, the output of the RF circuit becomes

$$v_o(t) = a_1 [V_1 \cos \omega_1 t + V_2 \cos \omega_2 t] \ldots \tag{7.27}$$

$$+ a_3 \frac{3}{4} [V_1^2 V_2 \cos(2\omega_1 - \omega_2)t + V_1 V_2^2 \cos(2\omega_2 - \omega_1)t] \ldots$$

---

3. The gain compresses because the sign of coefficient a_3 generally is opposite to that of coefficient a_1.

where those terms represent the signals at or near frequencies ω_1 and ω_2, as shown in Figure 7.8. The first-order term shows the ideal small-signal gain, and the third-order term at $2\omega_1 - \omega_2$, and $2\omega_2 - \omega_1$ corresponds to the third-order intermodulation response of the RF circuit. Although not shown, the other odd-ordered terms (due to coefficients a_5, a_7, and above) also create intermodulation distortion signals but with much smaller amplitudes.

The amplitudes of each amplified input signal V_{o1} and each third-order intermodulation product V_{o3} are described by

$$V_{o1} = a_1 V_i \qquad (7.28a)$$

$$V_{o3} = \frac{3}{4} a_3 V_i^3 \qquad (7.28b)$$

assuming V_1 and V_2 are equal to V_i. That leads to two important observations. First, the third-order intermodulation product grows much faster than the amplified input signal. Second, the extrapolated amplitudes of those two output signals eventually become equal at the third-order intercept point (IP3), as shown in Figure 7.9.

The relative amplitude of the third-order intermodulation product is given by

$$IM3 = \frac{|V_{o3}|}{|V_{o1}|} = \frac{3}{4} \frac{a_3}{a_1} V_i^2 \qquad (7.29)$$

while the actual output amplitude at the intercept point is found in the following way. The two output amplitudes, V_{o1} and V_{o3}, are equal when the input amplitude $V_i = V_{IP}$; therefore,

$$\frac{3}{4} a_3 = \frac{a_1}{V_{IP}^2} \qquad (7.30)$$

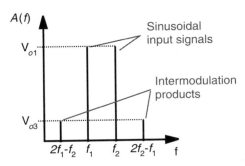

Figure 7.8 Distortion by a nonlinear amplifier creates intermodulation signals.

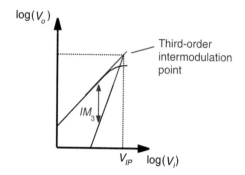

Figure 7.9 The IIP$_3$ occurs where the extrapolated desired and third-order responses intersect.

which when substituted into (7.28b) yields

$$V_{o3} = \left(\frac{a_1}{V_{IP}^2}\right) V_i^{\,3} \tag{7.31}$$

In a cascaded system like the one shown in Figure 7.10, the intermodulation distortion characteristics generally set the large-signal performance limit. If so, then the third-order intermodulation product at the output of the second stage is $a_{12}V_{o31} + V_{o32}$, which expands to

$$V_{o3} = a_{12}\left[\frac{a_{11}V_i^{\,3}}{(V_{IP1})^2}\right] + \frac{a_{12}V_{i2}^{\,3}}{(V_{IP2})^2} \tag{7.32}$$

where the input to the second stage V_{i2} equals $a_{11}V_i$. The amplified desired input signal is simply $V_{o1} = a_{11}a_{12}V_i$. When $V_i = V_{IP}$, the signals V_{o1} and V_{o3} are equal and thus

$$a_{11}a_{12}V_{IP} = \left[\frac{a_{11}a_{12}}{(V_{IP1})^2} + \frac{a_{11}^{\,3}a_{12}}{(V_{IP2})^2}\right]V_{IP}^{\,3} \tag{7.33}$$

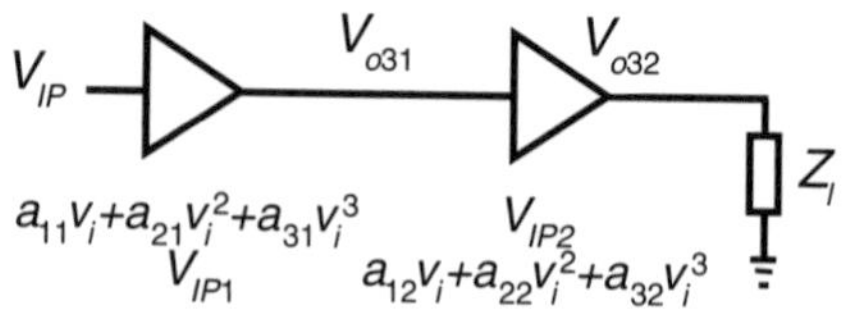

Figure 7.10 Cascaded intermodulation distortion.

which simplifies to

$$\left(\frac{1}{V_{IP}}\right)^2 = \left(\frac{1}{V_{IP1}}\right)^2 + \left(\frac{a_{11}}{V_{IP2}}\right)^2 \tag{7.34}$$

It is straightforward to extend (7.34) to an n-stage system:

$$\left(\frac{1}{V_{IP}}\right)^2 = \left(\frac{1}{V_{IP1}}\right)^2 + \left(\frac{a_{11}}{V_{IP2}}\right)^2 + \cdots \left(\frac{\prod_{i=1}^{n-1} a_{1i}}{V_{IPn}}\right)^2 \tag{7.35}$$

It is important to note that the later stages have a profound effect on the input intercept. That is due to the gain of the earlier stages, which increases the amplitude of the signals applied to the later stages. This situation is exactly opposite to that found during the cascaded noise figure analysis, where the early stages in the chain had the biggest effect.

7.2 Frequency Synthesis

The frequency synthesizer supplies the LO signals used by the transmitter and the receiver for frequency translation. It typically generates a single LO signal in the ultra-high frequency band (UHF band 300 MHz to 3 GHz)[4] and two LO signals in the very high frequency band (VHF band 3 MHz to 300 MHz), as shown in Figure 7.11.

The carrier frequency of the transmitted and received signals is governed by strict requirements set by national and international governing agencies. To meet those requirements, precise LO signals are needed. The signals are generated by phase-locked loops (PLLs) [8–11], like the one illustrated in Figure 7.12. The PLL consists of a frequency reference, divide-by-M and divide-by-N counters, a phase detector, a loop filter, and a voltage-controlled oscillator (VCO). The PLL is a feedback control system that minimizes the phase detector's output signal, which corresponds to the difference between the divided reference and the divided synthesized output. As a result, the system produces an output signal at the frequency $(M/N)f_{ref}$.

---

4. A single LO signal produces a transmit intermediate frequency (IF) that differs from the receive IF. That keeps the cost of the RF transceiver low and can improve isolation between the transmitter and the receiver.

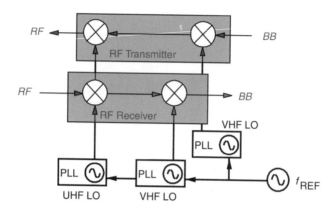

Figure 7.11 Frequency synthesizers are used by both the RF transmitter and the RF receiver.

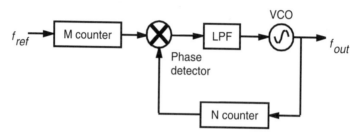

Figure 7.12 Block diagram of a PLL frequency synthesizer.

7.2.1 PLL Modes of Operation

The PLL has two modes of operation: acquisition and synchronization. In acquisition mode, the control loop moves the VCO's frequency toward the frequency of the reference signal. The process is highly nonlinear and characterized by the PLL's acquisition time, switching time, and pull-in range [12]. After acquisition, the PLL operates in synchronization mode, in which the VCO tracks the reference frequency. Frequency stability, accuracy, and spectral purity are key performance parameters in the synchronization mode of operation [12].

7.2.2 PLL Operation in Synchronous Mode

The operation of the PLL in synchronization mode is best analyzed in the phase domain using the simplified mathematical model in Figure 7.13 [11]. It consists of a phase detector, a low-pass filter, and a VCO. The phase detector compares the input time-dependent phase function $\Phi(t)$ to the system estimate

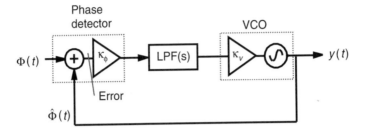

Figure 7.13 Linear approximation of the PLL.

$\hat{\Phi}(t)$, which includes noise $n_i(t)$. The gain of the loop drives the phase difference between the input and the output toward zero.

The phase detector is modeled by

$$v_{out}(t) = \kappa_\phi[\Phi(t) - \hat{\Phi}(t)] \qquad (7.36)$$

where κ_ϕ is the gain of the circuit, expressed in volts per rad. The loop filter conditions the system's response to the error signal and drives the VCO to produce an output signal equal to

$$y(t) = A\cos\left[\omega_o t + \kappa_\nu \int_t v_c(t)dt\right] \qquad (7.37)$$

where ω_o is the free-running frequency of the VCO, κ_ν is the gain of the VCO, and v_c is the control voltage. In (7.37), the important term is the phase component, because it is the loop feedback parameter. In the phase domain, the Laplace transform of the VCO output is

$$\hat{\Phi}_{out}(s) = \frac{\kappa_\nu}{s}V_c(s) \qquad (7.38)$$

The transfer function for the PLL is found using Mason's gain rule [13], which yields the closed-loop expression for the phase transfer function:

$$H(s) = \frac{\hat{\Phi}_{out}(s)}{\Phi_{in}(s)} = \frac{\kappa_\phi\kappa_\nu LPF(s)}{s + \kappa_\phi\kappa_\nu LPF(s)} \qquad (7.39)$$

Notice that the VCO adds a pole at dc to the transfer function. Consequently, a first-order loop filter $LPF(s)$ defined by

$$LPF(s) = \frac{1}{1 + \dfrac{s}{\omega_{LPF}}} \qquad (7.40)$$

with corner frequency ω_{LPF}, produces a second-order transfer function $H(s)$.

The behavior of the PLL is analyzed using control theory. The closed-loop transfer function for the PLL with a first-order loop filter can be rewritten in the familiar form

$$H(s) = \frac{\omega_n^2}{s^2 + 2\zeta\omega_n s + \omega_n^2} \qquad (7.41)$$

where the natural frequency ω_n is the gain-bandwidth product of the loop and ζ is the damping factor of the system, that is,

$$\omega_n = \sqrt{\omega_{LPF}\kappa_\phi\kappa_\nu} = \frac{1}{2}\sqrt{\frac{\omega_{LPF}}{\kappa_\phi\kappa_\nu}}$$

The stability of the system is analyzed using the open-loop magnitude-phase plot shown in Figure 7.14 for a typical second-order system. A well-designed system has a damping factor ζ greater than 0.5 and typically equal to $1/\sqrt{2}$.

The phase error transfer function is described by

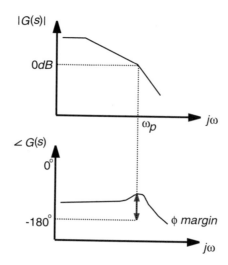

Figure 7.14 Open-loop response of PLL.

$$H_e(s) = \frac{\Phi_e(s)}{\Phi_{in}(s)} = \frac{s^2 + 2\zeta\omega_n s}{s^2 + 2\zeta\omega_n s + \omega_n^2} \qquad (7.42)$$

Note that the steady-state phase error is found when

$$\Phi_e(s) = \lim_{s \to 0} sH_e(s) \qquad (7.43)$$

and vanishes as s approaches zero. That demonstrates that the feedback control loop corrects the phase error of the VCO during steady-state operation.

7.2.3 PLL Nonideal Effects

The steady-state VCO output of the PLL is given by

$$y(t) = A\cos[\omega_o t + \hat{\Phi}_n(t)] \qquad (7.44)$$

where $\hat{\Phi}_n(t)$ is a noise process that is typically modeled as a zero-mean and stationary random variable. In the frequency domain, $\hat{\Phi}_n(t)$ describes the noise introduced by the loop and produces modulation or noiselike variations in the carrier.[5]

The N-counter of Figure 7.12 uses several cycles of the VCO signal to generate a comparison signal. That raises the noise level at the phase detector input by an amount equal to $20\log(N)$ [11]. In much the same way, the M-counter *lowers* the noise level of the reference by a factor of $20\log(M)$. If the noise generated by the phase detector is lower than either of its input noise levels, then the PLL output noise level is approximately $20\log(N/M)$ higher than the noise floor of the reference oscillator.

The VCO is another source of circuit noise in the PLL. Its noise psd can be estimated by [14]

$$S_\phi(\omega) = \frac{1}{4Q^2}\left(\frac{\omega_c}{\Delta\omega}\right)^2 \frac{P_n}{P_c} \qquad (7.45)$$

where $S_\phi(\omega)$ is the psd of the output, Q is the quality factor of the resonant tank circuit, ω_c is the frequency of oscillation, $\Delta\omega$ is the distance from the oscillation frequency, P_n is the noise power in the signal, and P_c is the strength of the carrier. To minimize phase noise, the Q factor and the carrier power must be as high as possible.

---

5. The carrier is the ideal single-frequency output signal.

The PLL's feedback reduces the phase noise of the VCO output near the carrier toward the level of the frequency reference, that is,

$$S_{PLL}(\omega) = S_\phi(\omega)H_e(s) \qquad (7.46)$$

That holds true as long as the loop can correct for phase errors. However, outside the loop filter bandwidth (frequencies $\Delta\omega$ higher than the filter bandwidth), the loop becomes ineffective, and the output noise level increases to that of the VCO, as shown in Figure 7.15.

The phase noise $S_\phi(\omega)$ near the carrier includes $1/f$ noise upconverted by the VCO's transistor amplifier and falls off at 9 dB/octave. The phase noise decreases at a rate of 6 dB/octave above the $1/f$ noise corner and up to the loop bandwidth of the PLL. Outside the bandwidth of the loop filter, the phase noise depends mostly on the VCO and remains relatively flat.

The output of the PLL also contains spurious tones due to modulation of the VCO control voltage at the comparison frequency. Those spurs are offset about the carrier (f_c) at

$$f_{spur} = f_c \pm Nf_{comp} \qquad (7.47)$$

where $f_{comp} = f_{ref}/M = f_{vco}/N$. Only spurs inside the loop bandwidth are attenuated, while others are unattenuated and included in the PLL output spectrum.

The loop filter plays a critical role in the performance of the PLL. A wide bandwidth accelerates acquisition and ensures low phase noise further away from the carrier frequency. A narrow bandwidth tolerates larger disturbances in the PLL and thereby maintains better synchronization. In most practical cases, the bandwidth of the loop filter is set to approximately one-tenth the reference frequency [15].

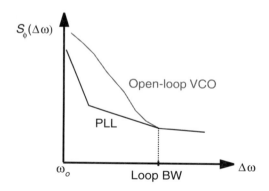

Figure 7.15 Phase noise performance of a PLL.

7.3 Transmitter System

The RF transmitter modulates the radio carrier with the baseband coded data and amplifies the resulting modulated waveform to the directed power level. In CDMA communication systems, the transmitter also includes precise power control and low-interference characteristics to permit a high number of users [15]. Those features add to the burden of portable operation and make transmitter design especially challenging.

The RF transmitter employs a two-step heterodyne architecture, consisting of an I/Q modulator, a variable gain amplifier (VGA), an RF mixer, a receive-band filter, a driver, a power amplifier (PA), and an isolator, as shown in Figure 7.16. The I/Q modulator performs the first frequency-translation step by superimposing the baseband data onto an IF carrier. The second frequency-translation step occurs in the RF mixer, which shifts the IF carrier to the prescribed radio frequency. A filter removes spurious products and lowers the system's noise floor.[6] The driver and the PA increase the strength of the modulated signal and couple it to the antenna. An isolator protects the PA against changes in the antenna's impedance. Those changes occur because of variations in the electric and magnetic (E&M) radiation patterns surrounding the antenna. For example, objects near the antenna, including the user's head, alter the impedance of the antenna.

The VGA provides power control in the transmitter. The circuit adjusts the transmit power from a low level of −50 dBm to a high level of +23 dBm (200 mW). The maximum level is low compared to other wireless standards, for example, the maximum power level for GSM is 2W [16]. Surprisingly,

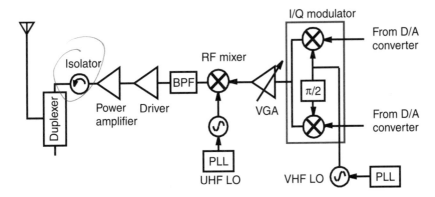

Figure 7.16 Block diagram of an RF transmitter.

---

6. It is vital to reduce the noise in the receive band because it appears at the antenna and is subsequently received by the RF receiver.

that does not translate to lower battery drain, because CDMA IS95 systems use a modulation scheme that requires linear and inefficient power amplification, whereas GSM systems use a constant envelope modulation technique that allows efficient PA operation.

7.3.1 Spurious Response

The ideal mixer produces an output signal whose amplitude is proportional to the input signal's amplitude and independent of the LO signal's level. That is because the LO signal does not have any information in its amplitude. As a result, the ideal mixer's amplitude response is linear for one input and independent of the other input. To make the mixer insensitive to the level of the LO signal requires a large LO signal level, which also leads to other benefits (which are discussed in Chapter 8).

Ideally, the mixer produces an output at frequency $f_{out} = f_{RF} \pm f_{LO}$, where f_{RF} is the input signal and f_{LO} is the LO signal. However, the nonlinear effects of the mixer distort the input signal and the large LO signal. As a result, the mixer produces output signals at frequencies described by

$$f_{M,N} = |Mf_{RF} \pm Nf_{LO}| \qquad (7.48)$$

where $f_{M,N}$ are spurious products, or spurs, with M and N integers ranging from $-\infty$ to $+\infty$. In practice, the amplitude of the spurious output signals decreases as N or M increases.

A spur table, like Table 7.1, lists the relative amplitudes of the output signals for a typical mixer. Of interest are the desired mixer signals, indicated by the $(M,N) = (1,1)$ products, a dc term associated with the $(0,0)$ spur, and the half-IF spur described by the $(2,2)$ spur. Note that these spurious signals propagate through the transmitter and can corrupt the modulated waveform[7] or interfere with the receiver.

7.3.2 Spectral Regrowth

Radio spectrum efficiency is a key parameter of wireless communication systems. To maximize efficiency, the mobile radio includes pulse-shaping filters in the digital modulator and very linear circuits in the RF transmitter. In practice, the transmitter displays nonlinear effects that spread the frequency spectrum of the modulated signal to nearby channels, as shown in Figure 7.17. This

---

7. The corruption occurs when the spurious signal mixes in a later nonlinear circuit and folds back to the modulated bandwidth.

Table 7.1
Spur Table for a Typical Mixer

M	N	Spurious Level (dB)
0	0	-30
1	1	0
1	2	-50
1	3	-40
2	1	-50
2	2	-50
2	3	-55
3	1	-40
3	2	-55
3	3	-55
4	1	-70
4	2	-75
4	3	-75
5	1	-65
5	2	-75
5	3	-70

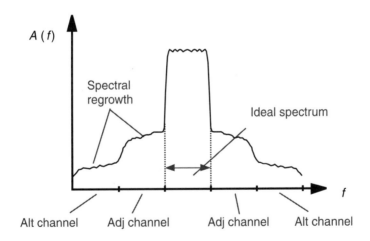

Figure 7.17 Spectral regrowth of transmitted signal.

phenomenon is referred to as spectral regrowth and is detrimental to system capacity.

Spectral regrowth is generated by intermodulation distortion between the signal components that make up the spread-spectrum direct-sequence modulated signal. Consequently, it depends on the statistical distribution of those signal components.

The digital modulator implements the modulation scheme and baseband filtering that shapes the amplitude variation of the modulated signal's envelope, as shown in Figure 7.18. One way of describing that variation is with the crest factor, which is defined by

$$\alpha = \frac{V_{pk}}{V_{rms}} \qquad (7.49)$$

where V_{pk} is the peak amplitude and V_{rms} is the rms value of the modulated waveform. For QPSK modulation, the crest factor is 6 dB, while for OQPSK modulation, the crest factor is 5.1 dB. The peak signal components are more likely to generate intermodulation distortion because they "drive" the transmitter circuits harder. To maintain acceptable linearity, the PAs are designed to operate linearly, even at the peak output power levels, which is responsible for the poor efficiency of the driver and PA circuits.

The best way to gauge spectral regrowth is by using a CDMA spread-spectrum input signal fed into the nonlinear amplifier. That is difficult to simulate, so an approximation based on third-order intermodulation distortion (IMD$_3$) is typically used [17, 18]. The intermodulation level is reduced by an empirical factor of 2 to 6 dB and normalized for bandwidth effects.

7.3.3 Noise

Noise also compromises the radio spectrum efficiency of wireless communication systems. That is because it adds to the background interference, lowering the signal-to-interference ratio (S/I), and it limits the useful power control range, raising the minimum transmit power level. Noise generated by the transmitter also affects the performance of the mobile radio's receiver. It raises the noise floor and thus increases the minimum signal level detectable by the receiver.

An ideal transmitter lowers the noise floor as it reduces the output power level and thereby maintains the SNR of the modulated signal. In practice, the

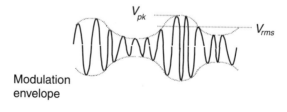

Figure 7.18 Illustration of peak and rms waveforms influencing the crest factor.

noise floor changes more slowly and eventually flattens out. If the signal level falls below the noise floor, the waveform quality and subsequent signal detection become poor.

Full duplex operation requires nearly complete isolation between the transmitter and the receiver. That is because any noise generated by the transmitter in the receive band elevates the background noise received at the antenna and degrades the sensitivity of the receiver, that is, its minimum detectable signal level. The received thermal noise floor is the available noise power, kTB. In a full duplex system, like the one shown in Figure 7.19, the noise power level at the receiver input N_{Rx} is

$$N_{Rx} = \frac{kTB + N_{Tx}}{L_{Duplex}} \tag{7.50}$$

where N_{Tx} is the output noise from the transmitter, kT is -174 dBm/Hz at 27°C, and L_{Duplex} is the loss through the duplex filter. That means the output noise from the transmitter needs to be less than -120 dBm/Hz in typical systems.[8]

The noise generated by the transmitter (N_{Tx}) is computed using noise factor (F) parameters for the circuits and the cascade analysis techniques outlined earlier. It is minimized by low-noise design techniques and filters that reject receive-band signals.

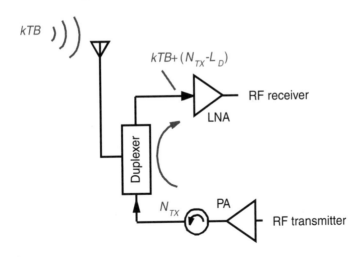

Figure 7.19 Noise level at the receiver input in an FDD system.

---

8. In practice, the transmitter-generated noise in the receive band should not raise the received noise level more than 1–2 dB.

7.3.4 Gain Distribution

The RF transmitter adjusts the output power over a wide range, which covers three particular challenging levels: maximum output power, critical output power, and minimum output power. At the maximum power level, the CDMA system demands low spectral regrowth and low receive-band noise, while minimizing battery drain. CDMA IS95 limits the adjacent channel power (ACP) in a 30-kHz bandwidth to -42 dBc or -60 dBm, whichever is larger. The first limit is a relative value ($ACP = P_0 - 42$ dB), while the second limit is an absolute power value. The two limits intersect at the critical output power level (-18 dBm), as shown in Figure 7.20. At the critical power level, the CDMA system demands both low spectral regrowth and low-noise performance. At the low-power level, only the output noise level remains important, and that becomes easy to meet because of the low-noise demands at the critical power level.

It is important to note that the relative limit of 42 dBc is essentially a restriction on the output noise floor. That is because as the output power level is decreased, the internal signals are made smaller and smaller, and thus the associated spectral regrowth is dramatically reduced. Eventually, the spectral regrowth falls below the noise floor, and the noise floor becomes the adjacent channel power limit.

The distribution of gain in the transmitter is crucial. Any noise generated by baseband and IF circuits is amplified by the RF circuits. That argues for keeping the RF gain low, which requires larger IF signal levels. However, the larger IF signals generate more distortion in the RF circuits. To minimize RF

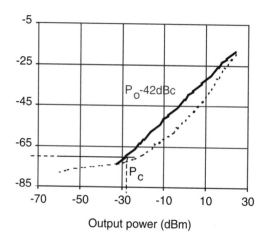

Figure 7.20 Graph defining critical power level for CDMA IS95.

distortion (and spectral regrowth), smaller IF signals and thus higher RF gain are needed.

The key parameters for the transmitter are output power, spectral regrowth, and noise level. Those parameters are charted in level diagrams that describe the signal-to-interference ratio of the modulated signal at different points in the transmitter. The level diagrams for the typical transmitter described in Table 7.2 are shown in Figure 7.21.

7.4 Receiver System

A communications radio receives and processes a variety of signals with a wide range of power levels. The system provides the necessary gain to extract very weak desired signals from various interfering signals. As such, the receiver is characterized by its ability to detect those weak signals, known as sensitivity, and its ability to reject strong interfering signals, known as selectivity.

The RF receiver employs a two-step heterodyne architecture consisting of a switched-gain low-noise amplifier (LNA), image reject filter (IRF), RF mixer, IF filter, VGA, I/Q demodulator, and low-pass filter, as shown in Figure 7.22. The LNA provides high gain for minimum cascaded noise figure and switches to low gain for strong input signal levels. The image reject filter removes the image signal and minimizes image noise. The RF mixer shifts the selected RF carrier to a common IF frequency using a variable-frequency LO. The I/Q demodulator then shifts the carrier frequency from IF frequency to baseband and separates the received signal into its I and Q components.

A VGA adjusts the level of the received signal to properly load the A/D converters. That maximizes the system's SNR. An IF filter, with a passband equal to the CDMA IS95 channel bandwidth of approximately 1.25 MHz, attenuates adjacent and nearby interfering signals. Low-pass filters at baseband

Table 7.2
Gain Distribution of the Transmitter

Function	Gain (dB)	NF (dB)	IP3 (dBm)
I/Q modulator	$P_o = -35$ dBm	45	0
VGA	−50 to 25	55 to 10	0 to +15
RF mixer	2	7	15
RF filter	−3	3	∞
Driver	15	3	20
PA	24	3	ACP = −44 dBc

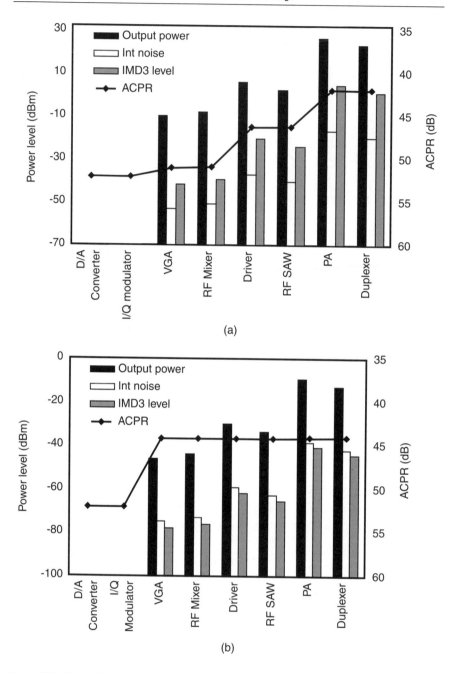

Figure 7.21 Level diagrams for RF transmitter: (a) maximum output power level and (b) critical power level.

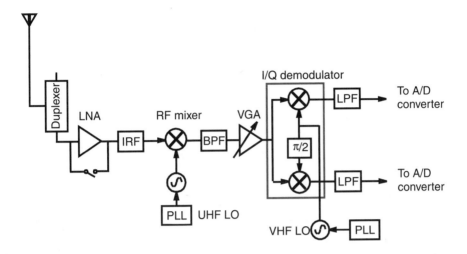

Figure 7.22 Block diagram of an RF receiver.

provide additional rejection of interfering signals. Together, the filters determine the receiver selectivity.

The image signal is a result of the downconversion mixing operation. The RF mixer shifts the entire spectrum of RF input signals (f_{RF}) down by the LO frequency according to the expression

$$f_{out} = |f_{LO} \pm f_{RF}| \qquad (7.51)$$

where f_{out} is the output of the mixer and f_{LO} is the frequency of the variable LO. That means that two different RF input signals, $f_{LO} - f_{IF}$ and $f_{LO} + f_{IF}$, can mix to the common IF frequency (f_{IF}). By design, one of those signals is the selected (or desired) RF carrier and the other is the image signal. Even without an image signal, the noise present at the image frequency is shifted to the IF frequency and is added to the receiver noise.

The spectrum of signals at the input to the receiver in a wireless environment is overwhelming. It typically consists of interfering signals in the same radio band, orthogonally coded users in the same frequency channel, and leakage from the mobile radio's transmitted signal. As a result, the design of the RF receiver is a formidable task.

7.4.1 Sensitivity

The weakest signal that can be received with a given SNR is referred to as the minimum detectable signal (MDS) [19]. It is set by the input-referred noise produced by the receiver system and is equal to

$$MDS = kTF_T B \qquad (7.52)$$

where F_T is the cascaded noise factor and B is the bandwidth of the system. The MDS is related to the sensitivity of the receiver in the following way

$$Sensitivity = kTF_T B(SNR_{min}) \qquad (7.53)$$

where SNR_{min} is a measure of the minimum SNR required by the demodulator. In CDMA IS95 communication systems, the spread-spectrum processing gain lowers the SNR requirements to less than -16 dB.

7.4.2 Selectivity

Poor receiver selectivity can lower the sensitivity, or "desensitize" the radio. That is generally accepted when there are strong interfering signals, but the performance impact is limited to less than 3 dB. The interfering signals lower the gain, or mask the wanted signal, by spurious mixing, blocking, intermodulation distortion, and cross-modulation effects. The effects are outlined below.

Half-IF mixing [11] is associated with the (2,2) spur in single-ended downconverters. An interfering signal midway between the desired receive signal (f_{RF}) and the LO down-mixing frequency (f_{LO}) at ($f_{RF} + f_{LO}$)/2 produces a masking IF signal, as illustrated in Figure 7.23. This is known as the *half-IF* problem and is suppressed by filtering the mixer's input signals and by using balanced circuit approaches, which minimizes second-order distortion.

A second mixing process that desensitizes the receiver is reciprocal mixing, which is illustrated in Figure 7.24 [20]. A strong interfering signal close in frequency to the desired signal mixes with a noisy LO signal to produce a

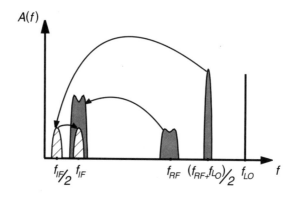

Figure 7.23 Half-IF problem in downconversion receiver.

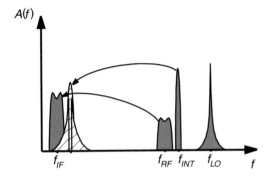

Figure 7.24 Reciprocal mixing.

masking signal. To prevent that, the phase noise of the frequency synthesizer must be kept lower than the relative strength of the interfering signal.

Blocking is another effect that degrades receiver sensitivity. It occurs when a strong interfering signal drives the receiver's circuits into compression and consequently lowers the gain applied to the wanted signal.

If an input signal consisting of a small desired component $V_1 \cos \omega_1 t$ and a very large undesired term $V_2 \cos \omega_2 t$ is applied to an RF system, modeled by (7.19), its output is of the form

$$v_o(t) = a_1 V_1 \cos \omega_1 t \ldots + \frac{a_3}{2}(V_1^3 + 3V_1 V_2^2) \cos \omega_1 t \ldots \quad (7.54)$$

where only the $\cos \omega_1 t$ terms are considered. Because V_1 is very small compared to V_2, the output signal is rewritten as

$$v_o(t) = V_o \cos \omega_1 t \approx a_1 V_1 \left(1 + \frac{3}{2}\frac{a_3}{a_1}V_2^2\right) \cos \omega_1 t \quad (7.55)$$

where V_o is the amplitude of the desired signal. In most cases, a_3 is opposite in sign to a_1, and so the gain decreases as V_2 increases.

In practice, the receiver's tolerance to a blocking signal is found by measuring the blocking signal's effect on the SNR. The desired signal is raised 3 dB above the receiver sensitivity, and a blocking signal is applied. The large interfering signal compresses the wanted signal, such that its voltage gain is

$$A_v = \frac{V_o}{V_1} = a_1 \left(1 + \frac{3}{2}\frac{a_3}{a_1}V_2^2\right) \quad (7.56)$$

A 3-dB decrease in gain occurs when

$$V_2 = 0.442\sqrt{\left|\frac{a_1}{a_3}\right|}\qquad(7.57)$$

which corresponds to a 3-dB decrease in SNR and sensitivity.

Low frequency and $1/f$ noise can also desensitize the receiver [20]. That is possible if there is an interfering signal at a frequency near the desired signal's frequency. The interfering signal mixes with the low-frequency noise and shifts its spectrum to the frequency of the interfering signal, as shown in Figure 7.25. The frequency-translated noise spectrum overlaps the desired signal and degrades performance.

In that situation, the input signal consists of the wanted signal $V_1 \cos\omega_1 t$, the interfering signal $V_2\cos\omega_2 t$, and the low-frequency noise modeled as $V_3\cos\omega_3 t$. The output of the RF system is

$$v_o(t) = a_1 V_1 \cos\omega_1 t \ldots + a_2 V_2 V_3 \cos(\omega_2 \pm \omega_3)t \ldots\qquad(7.58)$$

where a_2 is the second-order power series coefficient and V_3 is the noise level. Note that $\omega_1 \approx \omega_2$ and ω_3 is very small; therefore, $\omega_3 = \omega_1 - \omega_2$ and

$$v_o(t) = (a_1 V_1 + a_2 V_2 V_3)\cos\omega_1 t\qquad(7.59)$$

The actual gain is then

$$A_v = a_1\left(1 + \frac{a_2}{a_1}\frac{V_2 V_3}{V_1}\right)\qquad(7.60)$$

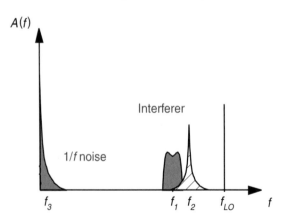

Figure 7.25 Frequency translation of low-frequency noise.

where a_2 is opposite in sign to a_1 in most cases. A 3-dB loss in SNR occurs when

$$V_2 V_3 = 0.292 \left| \frac{a_1}{a_2} \right| V_1 \qquad (7.61)$$

Intermodulation distortion is another effect that degrades receiver performance. It occurs when the harmonic distortion products of two strong interfering signals mix together, as shown in Figure 7.26. The input signal now consists of the desired signal $V_1 \cos \omega_1 t$ and two large interfering signals, $V_2 \cos \omega_2 t$ and $V_3 \cos \omega_3 t$. The effect of the RF system is to produce an output described by

$$v_o(t) = a_1 V_1 \cos \omega_1 t \ldots + a_3 \frac{3}{4} [V_2^2 V_3 \cos (2\omega_2 - \omega_3)t + \ldots] \qquad (7.62)$$

Assuming that $2\omega_2 - \omega_3$ approximately equals ω_1 and $V_2 = V_3 = V_i$, then the output can be rewritten as

$$v_o = \left(a_1 V_1 + a_3 \frac{3}{4} V_i^3 \right) \cos \omega_1 t \qquad (7.63)$$

The actual gain of the RF circuit in the presence of two interfering signals is then

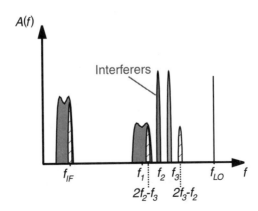

Figure 7.26 Intermodulation of two interfering signals in a receiver.

$$A_v = a_1 \left(1 + a_3 \frac{3}{4} \frac{a_3}{a_1} \frac{V_i^3}{V_1} \right) \qquad (7.64)$$

Again, a 3-dB impact in SNR is assessed, leading to

$$V_i = 0.730 \sqrt[3]{\left| \frac{a_1}{a_3} \right|} V_1 \qquad (7.65)$$

which shows the expected cubic power relationship.

The last mechanism that degrades receiver sensitivity is cross-modulation. Cross-modulation is a phenomenon in which the modulation of the strong transmit signal is transferred to a nearby interfering signal, as illustrated in Figure 7.27. If an input consisting of the desired signal $V_1 \cos \omega_1 t$, a nearby interfering signal $V_2 \cos \omega_2 t$ and the modulated transmit signal $V_3 [1 + m \cos \omega_m t] \cos \omega_3 t$, is applied to an RF circuit, its output can be described by

$$v_o \cong a_1 V_1 \cos \omega_1 t \ldots + 3 a_3 V_2 V_3^2 [1 + m \cos \omega_m t]^2 \cos^2 \omega_3 t \cos \omega_2 t \ldots$$
$$(7.66)$$

where m and ω_m are the modulation parameters of the transmit signal and only the $\cos \omega_1 t$ and the cross-modulation terms are considered. The output can be rewritten as

$$v_o \cong V_1 a_1 \cos(\omega_1 t) + 3 V_2 a_3 V_3^2 \cos(\omega_m t) \cos(\omega_2 t) \qquad (7.67)$$

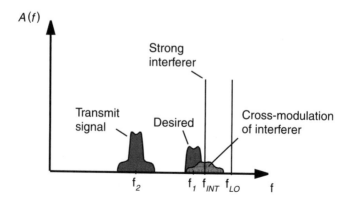

Figure 7.27 Cross-modulation effect.

which resembles the original modulated interfering signal, except that its carrier is located at ω_2 instead of ω_3. That shows that the modulation of the transmit signal is transferred to a nearby interfering signal.

The impact to SNR is analyzed as follows. The cross-modulation term compresses the desired signal while it raises the level of the noise or interference. This becomes a problem when the modulation bandwidth of the transmit signal, ω_m, is greater than the frequency separation between the desired signal, ω_1, and the interfering signal, ω_2, ($\omega_m \geq \omega_2 - \omega_1$). Under this condition, the cross-modulation signal overlaps the desired signal and thereby desensitizes the receiver.

In practice, there are several—perhaps dozens of—modulated interfering signals present at the receiving antenna. Those signals trigger any number of the described nonideal effects, which combine to reduce the receiver's selectivity.

7.4.3 Bit Error Rate and Frame Error Rate

Thus far, the performance of the receiver has been evaluated using SNR as the figure of merit. That is acceptable for analog communications, but digital communications are more accurately described by the probability of detection error, or bit error rate (BER). To do so requires an evaluation of the digital demodulator's performance.

The received spread-spectrum signal is despread by the digital demodulator during the recovery process. The process accumulates the energy in each message bit (E_b), giving

$$E_b \approx \frac{V_s^{\,2}}{\mathrm{Re}(Z)} T_b \qquad (7.68)$$

where V_s is the mean-square signal, $\mathrm{Re}(Z)$ is the real part of the load impedance and T_b is the period of each message bit. In contrast, the noise spectral density is

$$N_o \approx \frac{V_n^{\,2}}{\mathrm{Re}(Z)} T_c \qquad (7.69)$$

where V_n is the mean-square noise and T_c is the period of each chip. That results in the following SNR, as seen by the digital demodulator

$$\frac{E_b}{N_o} = \frac{V_s^{\,2}}{V_n^{\,2}} \frac{T_b}{T_c} = SNR \, \frac{W}{R} \qquad (7.70)$$

where E_b/N_o is the bit energy per noise spectral density, W is the spreading bandwidth (equal to $1/T_c$), R is the message bit rate (equal to $1/T_b$), and W/R is the spreading factor, or processing gain, of the spread-spectrum modulation.

The parameter E_b/N_o is critical because in digital communications that use BPSK and QPSK modulation (see Section 3.3.6), the probability of error is described by

$$P_e = Q\left[\sqrt{\frac{2E_b}{N_o}}\right] \tag{7.71}$$

which is used to analyze the BER. Note that noise in (7.71) refers to *any* unwanted energy that muddles the decision process. It includes unfiltered interfering signals, distortion produced by the receiver, and noise generated by the receiver.

One of the advantages of digital communications is the ability to digitally encode the source information, making it more tolerant of noise and interference. As a result, channel coding, such as convolutional codes and CRCs, further reduces the effective BER. The new error rate, called the frame error rate (FER), is measured after error correction.

To evaluate the RF receiver and digital demodulator requires simulation of radio propagation effects, but those effects are random and unpredictable (see Section 1.3). Thus, to measure FER, countless frames of data are transmitted, received, and demodulated.

7.4.4 Gain Distribution

The radio receiver must adapt to the different power levels while being sensitive to weak wanted signals and rejecting strong interfering signals. But that sets up contradictory requirements. To achieve low system noise figure and good sensitivity, high RF gain is needed. To minimize nonlinear effects that degrade selectivity, low RF gain is preferred.

Table 7.3 describes a typical receiver. The key parameters in that or any receiver are the wanted signal power, the noise power, the distortion power, and the SNR. The parameters are charted in level diagrams that illustrate the receiver's sensitivity and selectivity (Figure 7.28).

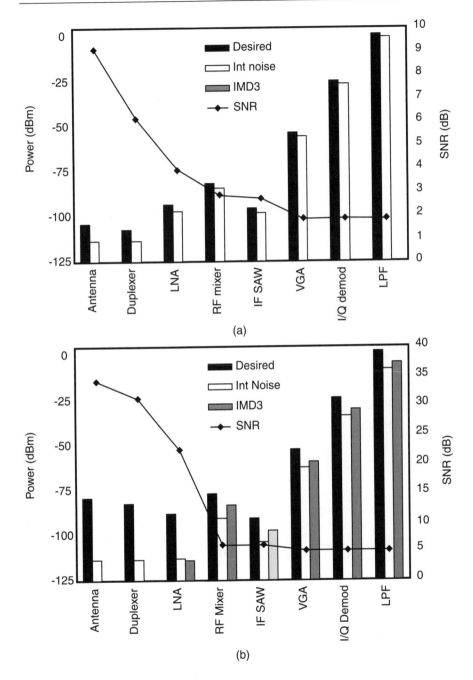

Figure 7.28 Level diagrams for the RF receiver: (a) sensitivity (low desired signal level) and (b) selectivity (high interfering signal levels).

Table 7.3
Gain Distribution of the Receiver

Function	Gain (dB)	NF (dB)	IP3 (dBm)
LNA	15	2.2	12
RF mixer	10	10	15
IF filter	−12.5	12.5	∞
Variable gain amplifier	−40 to 40	45 to 7	0 to −45
I/Q demodulator	30	25	25
Low-pass filter	26	75	20

References

[1]   Rappaport, T. S., *Wireless Communications: Principles and Practice*, Upper Saddle River, NJ: Prentice Hall, 1996.

[2]   Sklar, B., *Digital Communications*, Englewood Cliffs, NJ: Prentice Hall, 1988.

[3]   Couch, L. W., *Digital and Analog Communication Systems*, Upper Saddle River, NJ: Prentice Hall, 1997.

[4]   Taub, H., and D. L. Schilling, *Principles of Communication Systems*, New York: McGraw-Hill, 1986.

[5]   Gray, P. R., and R. G. Meyer, *Analysis and Design of Analog Integrated Circuits*, New York: Wiley, 1977.

[6]   Motchenbacher, C. D., and J. A. Connelly, *Low-Noise Electronic System Design*, New York: McGraw-Hill, 1993.

[7]   Friis, H. T., "Noise Figures for Radio Receivers," *IRE Proc.*, Vol. 32, July 1944, p. 419.

[8]   Crawford, J. A., *Frequency Synthesizer Design Handbook*, Norwood, MA: Artech House, 1994.

[9]   Best, R. E., *Phase-Locked Loops—Theory, Design, and Applications*, New York: McGraw-Hill, 1993.

[10] Larson, L. E., *RF and Microwave Circuit Design for Wireless Applications*, Norwood, MA: Artech House, 1997.

[11] Razavi, B., *RF Microelectronics*, Upper Saddle River, NJ: Prentice Hall, 1998.

[12] Lindsey, W. C., and M. K. Simon, *Phase-Locked Loops and Their Application*, New York: IEEE Press, 1977, pp. 1–7.

[13] Mason, S. J., "Feedback Theory: Some Properties of Signal Flow Graphs," *IRE Proc.*, Vol. 41, Sept. 1953.

[14] Leeson, D. B., "A Simple Model of Feedack Oscillator Noise Spectrum," *IEEE Proc.*, Feb. 1966, pp. 329–330.

[15] Salmasi, A., and K. S. Gilhousen, "On the System Design Aspects of Code Division Multiple Access (CDMA) Applied to Digital Cellular and Personal Communication Networks," *Proc. of IEEE Vehicular Technology Conf.*, VTC-91, May 1991, pp. 57–63.

[16]   Mehrotra, A., *GSM System Engineering*, Norwood, MA: Artech House, 1997.

[17]   Struble, W., et al., "Understanding Linearity in Wireless Communication Amplifiers," *IEEE J. of Solid-State Circuits*, Vol. 32, No. 9, Sept. 1997, pp. 1310–1317.

[18]   Kundert, K. S., "Introduction to RF Simulation and Its Application," *IEEE J. of Solid-State Circuits*, Vol. 34, No. 9, Sept. 1999, pp. 1298–1319.

[19]   Rohde, U. L, J. Whitaker, and T. T. Bucher, *Communications Receivers*, New York: McGraw-Hill, 1997.

[20]   Meyer, R. G., and A. K. Wong, "Blocking and Desensitization in RF Amplifiers," *IEEE J. of Solid State Circuits*, Vol. 30, No. 8, Aug. 1995, pp. 944–946.

8

RF Transmitter Circuits

The purpose of the RF transmitter of the mobile radio is to faithfully translate the digitally modulated waveform into a format suitable for transmission to the base station. Although straightforward in principle, the design of the RF transmitter is complicated by a variety of factors; to a large extent, it greatly affects the cost and dc power dissipation of the mobile radio.

Figure 8.1 is a block diagram of the RF transmitter. It receives orthogonal signals from the digital modulator by way of two dedicated D/A converters. The signals are combined and translated from baseband to a fixed IF frequency by an I/Q modulator. The upconversion process is followed by a VGA, which sets the power level of the transmitted signal. Typically, a second upconverter mixer shifts the fixed IF frequency of the modulated signal to the variable final

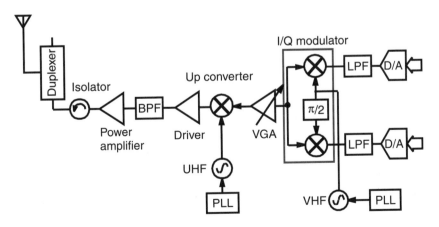

Figure 8.1 Block diagram of a typical CDMA RF transmitter.

187

frequency. A surface acoustic wave (SAW) filter follows the mixer to remove unwanted spurious signals and minimize out-of-band noise. Last, a PA amplifies the signal for transmission by the antenna.

This chapter presents the design techniques used for the RF transmitter functions, including the I/Q modulator, VGA, upconverter, SAW filter, and PA. Many of the functions overlap generically with those found in the RF receiver covered in Chapter 9. One of those circuits, the mixer, is introduced here and treated thoroughly in Chapter 9, because requirements on the RF receiver mixer typically are more demanding.

8.1 I/Q Modulator

The I/Q modulator is an efficient and convenient technique for generating phase-modulated signals.[1] Figure 8.2 is a block diagram of the I/Q modulator. It relies on two orthogonal signals, noted as I and Q, to produce a single complex waveform described by

$$s(t) = m_I(t)\cos(\omega_o t) + m_Q(t)\sin(\omega_o t) \qquad (8.1)$$

where $m_I(t)$ and $m_Q(t)$ are data sequences. Alternatively, (8.1) is sometimes expressed as

$$s(t) = Re\{A_c e^{j[\omega_0 t - \theta(t)]}\} \qquad (8.2)$$

where $A_c = \sqrt{m_I^2 + m_Q^2}$ and $\theta = \tan^{-1}(m_Q/m_I)$.

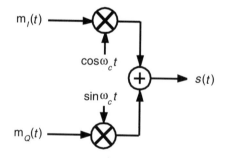

Figure 8.2 Block diagram of I/Q modulator used for generating a complex QPSK-modulated signal.

---

1. The concept of phase modulation using the I/Q modulator was introduced in Section 7.1.3.

QPSK modulation is used in CDMA IS95 systems as well as many other wireless communication systems. To ease RF linearity requirements, CDMA IS95 delays the Q data by one-half chip and thereby prevents simultaneous transitions of both orthogonal data streams. This is known as OQPSK. As a result, the trajectory, or path, of the signal in phase space stays clear of the origin, and—beneficially—the envelope of the modulated signal shows less variation. In fact, OQPSK modulation lowers the peak to the average ratio of the modulated signal by about 0.5 dB.

The I/Q modulator is a direct upconverter that transforms the frequency spectrum of each orthogonal input signal to the IF carrier frequency. Ideally, it produces a 1.23-MHz wideband spread-spectrum signal, suppresses the carrier signal, and preserves orthogonal signal relationships.

8.1.1 Nonideal Effects in the I/Q Modulator

In practice, the I/Q modulator is plagued by carrier leakage, I/Q leakage, AM-to-AM conversion, and AM-to-PM conversion [1]. Carrier leakage is caused when the input signals are dc offset. The reason for that is as follows. The I/Q modulator uses two mixers to translate each of the input signals to orthogonal waveforms at the IF carrier frequency. The output of each mixer is simply the product of an input signal and one of the orthogonal carrier signals. When an input signal has a dc offset, a portion of the carrier signal appears at (or leaks to) the output of the mixer.

A dc offset can be caused by circuit and device mismatches before the I/Q modulator as well as within the mixer circuits. Metal oxide semiconductor field effect transistor (MOSFET) circuits generally are preferred at baseband because of very low power consumption, but those transistor structures provide poor matching and thus larger dc offsets, typically 5–10 mV [2]. Better matching is available with bipolar transistors, typically less than 1 mV [3]. Still, to achieve very low carrier leakage, a dc offset correction scheme usually is employed [4].

I/Q leakage is due to phase and/or amplitude imbalance in the input signals or LO carrier signals. As a result, the outputs of the mixers are not orthogonal and actually corrupt, or spill into, each other. The leakage can be found using the following expression [5]:

$$\frac{P_{leakage}}{P_{desired}} = \frac{1 - 2\sqrt{\Delta A/A}\cos\Delta\theta + \Delta A/A}{1 + 2\sqrt{\Delta A/A}\cos\Delta\theta + \Delta A/A} \tag{8.3}$$

where $\Delta A/A$ is the power gain ratio and $\Delta\theta$ is the phase mismatch between the I and Q channels. In practice, I/Q leakage less than −25 dB is satisfactory.

The mixers that comprise the I/Q modulator also generate unwanted spurious products. These products can be attenuated by a simple inductor-capacitor (LC) filter, which provides about 15 dB of attenuation at three times the IF carrier frequency. Additionally, circuit techniques, such as fully differential circuits and feedback, can be used to combat even- and odd-order distortion.

8.1.2 I/Q Modulator Circuit Techniques

The gain of the two orthogonal mixers is matched by using common analog techniques that ensure equal input signal and LO carrier signal levels. These circuits typically are fabricated in close proximity and therefore exhibit very good gain matching. In practice, phase matching is more difficult. It requires truly orthogonal LO carrier signals, a lack of which leads to the major source of I/Q leakage.

A common technique to generate orthogonal signals is through lead/lag (high-pass/low-pass) filters. These filters are simple RC structures, as shown in Figure 8.3(a), that shift the LO carrier signal ±45 degrees. The amplitude mismatch and the phase mismatch of the structures are given by

$$\Delta A \approx \frac{1 - (\omega RC)^2}{1 + (\omega RC)^2} \tag{8.4}$$

and

$$\Delta \theta \approx \frac{\pi}{2} + \tan^{-1}(\omega R_2 C_2) - \tan^{-1}(-\omega R_1 C_1) \tag{8.5}$$

where R is the mean of the resistances and C is the mean of capacitances.

Another technique to generate orthogonal signals is the phase sequence asymmetric polyphase filter, shown in Figure 8.3(b). An extension of the lead/lag filters, it provides antisymmetric properties, rejects all nonquadrature components, and yields improved orthogonal signals [6]. In practice, two filter stages typically are cascaded to achieve improved performance [7].

Still another orthogonal technique uses a clock signal at two times the LO frequency. The signal is applied to a flip-flop, producing orthogonal signals at the output of each latch, as shown in Figure 8.3(c). The phase error is

$$\Delta \theta \approx \sqrt{2} \tan^{-1} \left(\frac{P_{2N}}{P_F} \right) \tag{8.6}$$

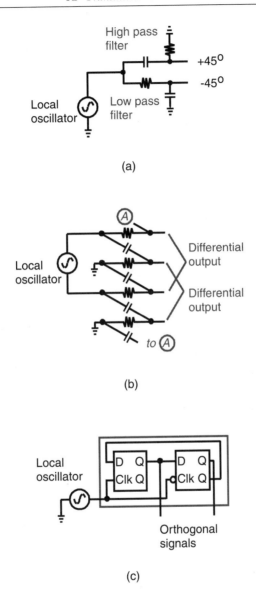

Figure 8.3 Schematic diagram of circuits for generating orthogonal LO signals for the I/Q modulator: (a) low-pass/high-pass structure, (b) phase-sequence asymmetric polyphase filter, and (c) digital technique using frequency dividers.

where the ratio (P_{2N}/P_F) is the relative level of the clock signal's even harmonics. If the second harmonic is suppressed 20 dB, the phase error is less than 1 degree. Odd harmonics do not affect the phase error.

In some applications, it is necessary to reduce the phase error further. One way to do that, Haven's technique [8], relies on vector processing, as diagrammed in Figure 8.4. With nearly identical amplitudes, the phase error is reduced to

$$\Delta\theta' = \tan^{-1}\left[\frac{2 - \Delta A/A}{\Delta A/A}\ \tan\left(\frac{\Delta\theta}{2}\right)\right] - \tan^{-1}\left[\frac{\Delta A/A}{2 - \Delta A/A}\ \tan\left(\frac{\Delta\theta}{2}\right)\right]$$

(8.7)

where $\Delta A/A$ is the amplitude mismatch and $\Delta\theta$ is the original phase difference.

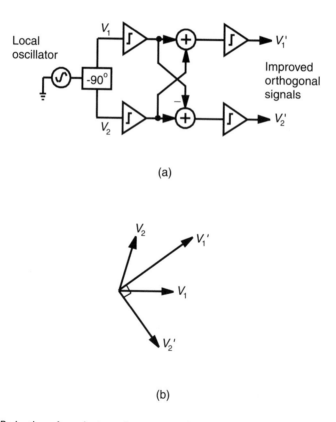

(a)

(b)

Figure 8.4 Reduction of quadrature phase errors through the use of Haven's technique: (a) block diagram, and (b) vector processing.

8.2 Power Control in the RF Transmitter

The VGA is an important function that allows power adjustment in the RF transmitter and, as such, provides a critical part of the power control algorithm. The requirements on the VGA exceed simple gain control; it also must limit spurious regrowth and noise, as outlined in Section 7.3.

The VGA relies on transistor-based circuits that are designed to adjust gain in a predictable manner. The control must be predictable because the RF transmitter operates "open loop," that is, without any feedback. That is because it is impractical to sense the output power of the RF transmitter over a window from −50 dBm to +23 dBm. For instance, diode detectors typically provide a dynamic range of only 25 to 30 dB.

The classical approach to gain control is shown in Figure 8.5(a). In that circuit, the input voltage signal is amplified and converted to a current signal by transistors Q_5 and Q_6 and bias current I_1. A portion of the current signal is steered by transistors Q_1 and Q_4 to the load resistors, where it is translated to an output voltage signal. It is also possible for a portion of the current signal to be diverted away from the load resistors using transistors Q_2 and Q_3. As a result, gain control is provided via the four steering transistors, Q_1 to Q_4. The gain of that circuit is approximately

$$A_\nu = \frac{g_m R_1}{1 + e^{-V_c/V_T}} \tag{8.8}$$

where g_m is the transconductance of transistors Q_5 and Q_6, V_C is the control voltage, and V_T is the thermal voltage (about 26 mV at room temperature). The gain control of the circuit is nonlinear, but it can be made linear with simple circuit techniques [9]. However, the circuit suffers several drawbacks. The tail current, power dissipation, input linearity, and noise level are all relatively constant, even as the gain is adjusted.

The VGA shown in Figure 8.5(b) improves on the classical amplifier circuit. Here, the bias current is adjusted based on the desired output power level. That alters the transconductance and gain, as well as the dc operating point, of the circuit. The gain of this circuit is simply

$$A_\nu = \frac{I}{2V_T} R \tag{8.9}$$

where I is the adjustable bias current and R is the load resistance. The input voltage range of this circuit is limited to less than $2V_T$, although that can be extended by using a multitanh input stage [10]. Note that emitter degeneration

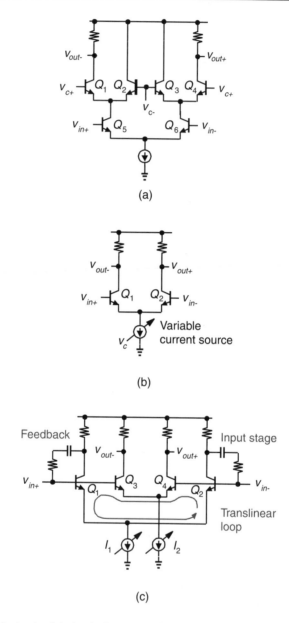

(a)

(b)

(c)

Figure 8.5 VGA circuits: (a) classical structure, (b) variable bias current, and (c) translinear loop.

also increases the useful input voltage range; however, this feedback technique stabilizes gain and therefore prevents adjusting gain via the bias current.

The translinear circuit shown in Figure 8.5(c) further improves the gain control circuit. It consists of a linearized input stage and a high-current output stage that are coupled using the translinear principle. The input stage consists of a differential pair (transistors Q_1 and Q_2) with resistive shunt feedback to stabilize gain. Linearizing gain also stabilizes the base-emitter voltages associated with transistors Q_1 and Q_2. That is critical because those junctions, along with the base-emitter junctions of transistors Q_3 and Q_4, form the translinear loop. Furthermore, the devices are well matched because of integrated circuit techniques. As a result, the output current is proportional to the input current to the amplifier. The gain of the translinear amplifier is, therefore,

$$A_i \approx \left(\frac{R_2}{R_1} + 1\right)\left(\frac{I_2}{I_1}\right) \qquad (8.10)$$

where the first term is the gain of the input stage and the second term is the effect of the translinear loop conveyor [11]. Note that bias current I_2 easily controls the gain of the circuit. Additionally, the linearity of the amplifier tends to increase with growing bias current.

8.3 Upconverter Design

The frequency upconversion process from the IF frequency to the final RF frequency is accomplished by a simple mixing operation, as shown in Figure 8.6. The input signal is multiplied by an LO operating at the appropriate frequency to produce the desired output. As with all multiplier circuits, the output signal consists of products at the sum and difference frequencies of the two input signals. In low-side injected mixers, the frequency of the LO is below the frequency of the desired output signal. Consequently, the desired mixer product is the sum term. By contrast, in high-side injected mixers, the

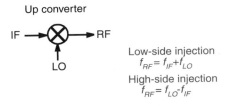

Figure 8.6 Simple mixing operation shifts carrier from IF to RF.

frequency of the LO signal is above the frequency of the desired mixer output signal, and the desired mixer product is the difference term.

The most important issue in the upconverter is linearity. The mixer inevitably will exhibit some nonlinearity, which can be characterized by its input or output third-order intercept point or, alternatively, by its adjacent channel power at a given output power level. The linearity of the mixer is crucial only at the RF port; nonlinearities at the LO port of the mixer are filtered away by the sharp response of subsequent filters. Another important consideration in the upconverter mixer is noise, both in-band and in the receiver band. At low power levels, it is important to suppress circuit noise and preserve the SNR of the transmitted signal. At high power levels, it is important to minimize noise in the receive band to avoid desensitization of the RF receiver.

The transistor level design of mixer circuits is presented in detail in Chapter 9.

8.4 SAW Filter Technology

In a typical wireless transceiver, filters perform the all-important roles of duplexing, image elimination, spurious rejection, and channel selection. Those devices represent the one area of radio design that still remains largely the province of passive hybrid techniques whose origins date back to the early days of radio communications. That is due in large part to the extreme dynamic range and energy storage requirements that they must meet, usually eliminating the possibility of active integrated circuit approaches. This section summarizes a few key filter parameters and details an extremely important class of filters used in transceivers, the SAW filter.

Figure 8.7 shows the "typical" magnitude response of a lowpass filter. It exhibits nonzero loss in the passband, as well as nonzero gain in the stopband. The shape factor of the filter is described by

$$Shape\ Factor = \frac{f_{Stop}}{f_{Pass}} \tag{8.11}$$

where f_{Stop} is the stopband or "skirt" bandwidth at some predetermined loss and f_{Pass} is the filter bandwidth at some predetermined gain. An ideal filter has a shape factor of unity, although values in the 1.5 to 3 range are considered excellent for most wireless applications.

The second factor of importance is group delay, a measure of phase linearity. Group delay is defined mathematically as

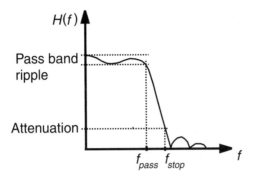

Figure 8.7 Typical filter response illustrating passband and stopband.

$$T_g(\omega) = \frac{\partial}{\partial \omega} \theta(\omega) \qquad (8.12)$$

where $\theta(\omega)$ is the phase response of the filter at frequency ω. As such, it indicates the phase distortion or smearing expected when a modulated signal passes through the filter. An ideal filter exhibits constant group delay and linear phase response. In general, the largest group delay variation occurs at the passband edge of the filter response.

SAW filters have characteristics that approach the properties of ideal filters. They exhibit outstanding linearity, extremely narrow transition bands, and very "flat" group delay characteristics, at the expense of rather high insertion loss and cost. In fact, the refinement of SAW devices made the development of low-cost modern digital communications possible [12].

SAW filters are attractive because they can be a direct physical implementation of a tapped delay-line FIR filter. They rely on the piezoelectric transduction of a surface acoustic wave through a crystal with significant piezoelectric activity, typically $LiNbO_3$ or quartz. Because the physical propagation through the medium is extraordinarily rapid, the filters can be made very compact. A simplified cross-sectional diagram of a typical SAW filter is shown in Figure 8.8.

The wave-generating and receiving transducers are fabricated as interleaved metallic (usually aluminum) combs deposited on the flat surface of a piezoelectric material. A sinusoidal voltage applied between the fingers of the input transducer creates a piezoelectrically induced acoustic wave running perpendicular to the fingers. When the waves appear under the receiving electrodes, they produce a voltage related to the material deformation. Like an array of antennas, the highest gain occurs at a frequency where the surface wavelength of the wave matches the spacing between the electrodes.

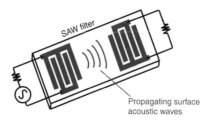

Figure 8.8 Cross-section of typical SAW filter.

SAW filters have a rather high insertion loss in the passband, typically from 3 to over 15 dB. That is because the surface acoustic waves travel in two directions. The waves encounter some loss through the material as they propagate, and the piezoelectric transduction exhibits some significant loss at each end of the filter. Nevertheless, their performance has improved dramatically in recent years—some SAW devices exhibit an insertion loss approaching 1 dB. The design of SAW filters is extremely advanced and highly specialized, but their basic operation can be analyzed with a simple example.

In a typical case, the acoustic wave travels from the input transducer to the output transducer with velocity v (the speed of sound in the material). The signal is received at the output transducer by the multiple N electrodes, whose spacing is l and whose area is proportional to a_n. Hence, the output signal is approximately

$$v_{out}(t) = \sum_{i=0}^{N-1} a_n v_{in}(t - nT) \tag{8.13}$$

where $T = vl$.

The filter is designed by appropriately weighting the coefficients a_n. In the simplest possible case, the areas of all of the electrodes are equal, and the resulting values of a_n can be set to unity. Therefore,

$$H(s) = \frac{v_{out}(s)}{v_{in}(s)} = \sum_{i=0}^{N-1} e^{-nTs} = \frac{1 - e^{-sNT}}{1 - e^{-sT}} = \frac{\sin\left(\dfrac{\omega TN}{2}\right)}{\sin\left(\dfrac{\omega T}{2}\right)} e^{-j(N-1)\omega T/2}$$

$$\tag{8.14}$$

Note that the resulting phase response of the filter is perfectly linear, and the ideal filter exhibits constant group delay. The magnitude response of the filter exhibits a periodic bandpass characteristic, as shown in Figure 8.9. The

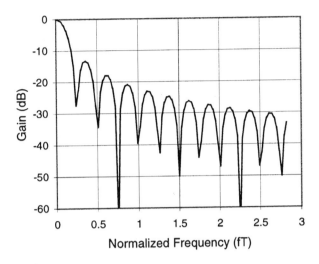

Figure 8.9 Frequency response of SAW filter with equal tap weighting.

width of the main lobe can be narrowed by increasing the number of electrodes N. Of course, the $\sin Nx/\sin x$ amplitude response is not necessarily ideal for most applications.[2] Weighting the taps by an appropriate amount can improve the amplitude response of the filter considerably.

The optimized weighting of the taps is accomplished by a variety of techniques. Apodization varies the physical size of each tap to vary the amplitude weighting of each tap. Finger withdrawal removes some fingers to provide phase weighting of the response in the time domain. Figure 8.10 illustrates both techniques. Furthermore, the aforementioned loss in the transducer, due to the bidirectional propagation of the acoustic wave across the material surface, can be minimized through the use of a symmetrically divided output transducer, as shown in Figure 8.11. In that case, the bidirectional wave is "captured" by multiple output transducers.

SAW devices are extremely sensitive to termination and ground-loop limitations, because they are required to provide extraordinarily high rejection of out-of-band signals. Figure 8.12(a) is an example of a typical ground-loop problem encountered with a SAW filter. One end of the input and output electrode are tied together through a common ground connection to an output pin, which encounters some small inductance before it reaches ground. In this case, the out-of-band rejection is then limited to approximately the ratio of the inductive reactance and the input impedance. For that reason, SAW devices are often configured in a differential mode at one or both ports, as shown in

---

2. The comb-filter response was also described in Section 3.3.2.

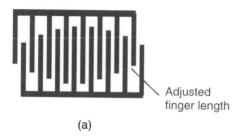

(a)

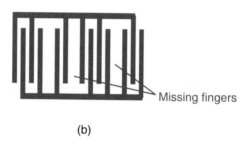

(b)

Figure 8.10 Optimized tap weighting of SAW filters: (a) apodization and (b) finger withdrawal.

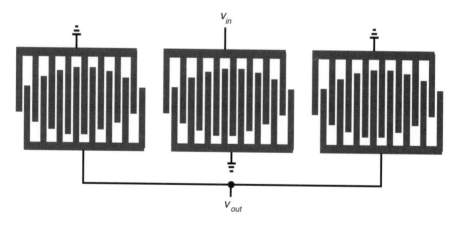

Figure 8.11 Use of symmetrically divided output transducer to reduce loss in SAW filter.

Figure 8.12 (b), to minimize that effect. Careful board layout techniques can also mitigate the problem to a reasonable extent.

8.5 Power Amplifiers for Transmitter Applications

Power amplifiers typically dissipate more dc power than any other circuit in the mobile radio. That is because the PA is ultimately responsible for closing

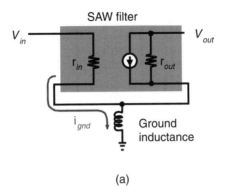

(a)

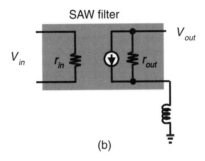

(b)

Figure 8.12 (a) Illustration of ground-loop problems with SAW filter board layout. (b) Improved out-of-band rejection can be achieved through the use of a balanced structure on one or both ports.

the link to the base station receiver. As such, it needs to be capable of transmitting the peak output power (200 mW for class III mobile radios), although the average transmitted power usually is considerably smaller (typically a few milliwatts [13]). Furthermore, the PA usually is designed for worst-case performance, making it difficult to reduce power consumption at lower transmit power levels.

The utility of the mobile radio depends on RF transmitter efficiency and, to a certain extent, on available battery technology. The energy limitations of traditional battery-powered mobile radios are significant and require careful planning. Li-Ion battery cells and traditional NiCd or NimH cells are used to power today's mobile radios. Those devices provide a nominal output voltage between 4.5V and 3V, depending on the charge state, as shown in the discharge curves in Figure 8.13.

Increasing the utility of the mobile radio requires major improvements in battery technology and PA design. Basic issues for battery technology include

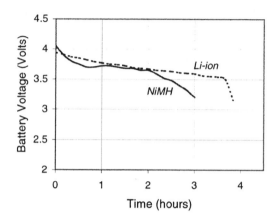

Figure 8.13 Comparative cell voltages of NiMH and Li-ion batteries during discharge.

cell size and energy storage density. Some key requirements and design considerations for CDMA PAs[3] are outlined next.

8.5.1 PA Design Specifications

The design of PAs is complicated by a variety of factors that make the simultaneous achievement of high performance and high efficiency difficult. Table 8.1 compares the specifications for a typical CDMA IS95 PA with other wireless communication standards.

Table 8.1
Comparison of PA Characteristics for Popular Wireless Communication Systems

Parameter	GSM	NADC	CDMA IS95
Frequency range (MHz)	890–915	825–849	825–849
	1,710–1,785	1,850–1,910	1,850–1,910
Maximum transmit power (dBm)	30.0	27.8	23.0
Long-term mean power (dBm)	21.0	23.0	10.0
Transmit duty cycle (%)	12.5	33.0	Varies
Occupied bandwidth (kHz)	200	33	1228
Modulation method	GMSK	$\pi/4$-QPSK	OQPSK
ACPR (dBc)	n/a	−26	−26
Peak-average ratio (dB)	0	3.2	5.1
Typical efficiency (%)	>50	>40	>30

3. These design issues are also valid for most wireless communication systems.

From the perspective of PA implementation, the key requirement on the RF transmitter is minimal spurious radiation at high output power levels. That covers spectral regrowth, also referred to as ACP, and any unwanted mixing products. A correspondence has been shown between spectral regrowth and intermodulation distortion, even though intermodulation distortion typically is a small-signal measurement. That relationship is [14]

$$IIP3 = -5\log\left[\frac{P_{IM3}(f_1, f_2)B^3}{P_0[(3B - f_1)^3 - (3B - f_2)^3]}\right] + 22.2 \qquad (8.15)$$

where $IIP3$ is the required input intercept point in dBm, B is one-half of the signal bandwidth, f_1 and f_2 are the out-of-band frequency limits, P_o is the output power of the amplifier, and $P_{IM3}(f_1, f_2)$ is the out-of-band specified power.

Another important consideration in the design of the PA for wireless applications is the level of out-of-band noise. In full-duplex operation, excessive noise in the receive band can corrupt the received signal and desensitize the receiver of the mobile radio. Fortunately, most PA transistors make acceptable low-noise amplifiers.

Although the peak dc-to-rf efficiency of the PA occurs at the peak output power, the PA itself rarely operates at that power level. That is illustrated by the transmit power probability profiles shown in Figure 8.14. As a result, it is extremely important to consider average efficiency when considering the optimum PA configuration. In this case, the average efficiency of a PA can be calculated as

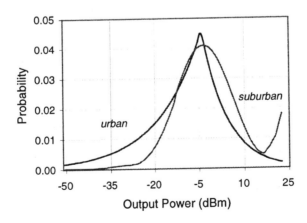

Figure 8.14 Probability curves for transmit power level in urban and suburban environments [15].

$$\eta_{avg} = \frac{\displaystyle\int_0^{P_{out_{max}}} P_{out}\, p\,(P_{out})\, dP_{out}}{\displaystyle\int_0^{P_{out_{max}}} P_{dc}\,(P_{out})\, p\,(P_{out})\, dP_{out}} \qquad (8.16)$$

where P_{out} is the output power, $p\,(P_{out})$ is the probability of the output power P_{out}, and $P_{dc}\,(P_{out})$ is the dc power required at P_{out}. In practice, this quantity is the measure of the effectiveness of the PA converting the battery stored energy into transmitted energy and is considerably less than 10%.

Finally, the PA is required to deal with the rugged physical environment of a typical mobile radio through its interface with the antenna. The problem arises when the antenna is suddenly grabbed or is too close to a conducting surface. In that case, the voltage standing wave ratio (VSWR) of the antenna can rise dramatically. In the worst case, the peak drain or collector voltage can rise to four times the dc power supply voltage [16]. To avoid potential disaster, most mobile radios include low-loss isolators in the transmit path, which effectively isolate the PA from any mismatch effects at the antenna port. That is possible because of technology developments from diverse fields.

The architecture of a mobile radio PA usually consists of several stages of gain, with the first few stages of amplification referred to as the driver stages, and the final stage referred to as the output stage. The overall gain of the circuit is in the 25- to 35-dB range, takes a signal at relatively modest power levels, and converts it to roughly 400 mW.[4] The design of the driver stages is quite straightforward, usually consisting of simple common-emitter or common-source amplifiers. The required linearity and efficiency of these stages are straightforward to achieve, and most of the design effort is focused on the output stage.

8.5.2 PA Design Techniques

Because of the stringent requirements on power efficiency and linearity, the typical PA designer has very little latitude in the design and implementation of the circuit. That can easily be seen from an analysis of a simple common-emitter PA, as shown in Figure 8.15(a). The amplifier is designed to deliver 400 mW, which has a nominal impedance of 50Ω. That will require an rms voltage at the collector of about 4.5V and a peak-to-peak voltage swing greater

---

4. Although the peak transmit power level is 200 mW, the isolator and the duplex filter attenuate the signal about 3 dB before reaching the antenna, thus making the PA work harder.

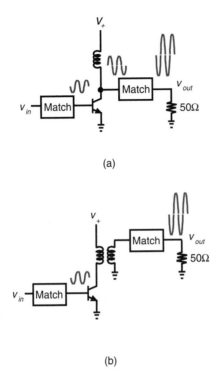

(a)

(b)

Figure 8.15 Common-emitter PA design: (a) with 50Ω load impedance and (b) with impedance transformer.

than 12V. The dc value of the collector must be at least half that value, or 6V.

Because the dc voltage of the collector typically is limited by the onboard power supply of 3 to 4.5V, there clearly is a mismatch between the conveniently available power supply and the amplifier requirements. Furthermore, modern high-frequency transistors suitable for this application typically have low break-down voltages. The problem is even more challenging when sudden impedance changes increase VSWR. For best operation, the transistor can deliver this level of power into a much lower impedance, typically 10Ω or less.

The solution to the problem lies in a lossless impedance transformation between the 50Ω load impedance and the much lower impedance required by the device, as shown in Figure 8.15(b). Now, the voltage swing is reduced by the square-root of the impedance transformation ratio, and the current is increased by the same amount. For all intents and purposes, the amplifier can be analyzed as if it is driving a much lower load impedance, a substantial advantage.

Since ideal transformers are impractically lossy at these frequencies, the impedance transformation typically is accomplished by a series of "L-matches" of progressively decreasing impedance, as shown in Figure 8.16. The loss through the network is decreased by using several stages, instead of a single stage. It is fairly easy to demonstrate that the overall loss of a two-stage network is minimized when the intermediate impedance of each stage is the geometric mean of the impedance at each end of the network [17].

The choice of optimum load impedance for the amplifier to achieve the output power level, linearity, and efficiency is difficult. It depends in part on the I-V curve of the power transistor. A typical curve is shown in Figure 8.17. In that case, the maximum collector current is given by I_{max}, and the maximum collector voltage—prior to breakdown—is given by V_{max}. Clearly, the largest power delivered to the load impedance occurs when the transistor reaches both those limits. In that case, the peak-to-peak voltage swing at the load is V_{max}, and the peak-to-peak current swing at the load is I_{max}. That produces an rms power at the load of

$$P_{load} = \frac{V_{max}}{2\sqrt{2}} \frac{I_{max}}{2\sqrt{2}} = \frac{V_{max} I_{max}}{8} \qquad (8.17)$$

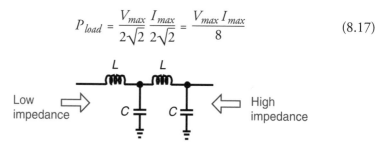

Figure 8.16 Impedance matching network.

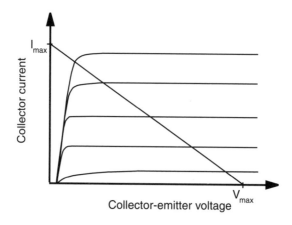

Figure 8.17 Power transistor I-V curve, showing location of ideal load line.

and the load impedance presented to the device to extract that power is given by

$$R_{load} = \frac{V_{max}}{I_{max}} \qquad (8.18)$$

When the device is operating in this regime, it is in the class A mode of operation, and the conduction angle is said to be 360 degrees, confirming the fact that the device is "on" during its complete operating cycle. For example, a 400-mW amplifier will require a load resistance of 10Ω when the device is limited to 6V-peak collector excursion.

The dc power dissipation and resultant efficiency are also important considerations in the design of a radio frequency PA. That is typically assessed by measuring the power-added efficiency of the device, which is defined as

$$\eta_{PAE} = \frac{P_{rf(out)} - P_{rf(in)}}{P_{dc}} \qquad (8.19)$$

where $P_{rf(in)}$ is the RF input power, $P_{rf(out)}$ is the output power, and P_{dc} is the dc power dissipation of the amplifier. In the low-frequency limit of operation, the gain of the amplifier is very high, and there is no RF input power. Then the expression for the efficiency is simply the ratio of the RF output power to the dc input power, often called the collector, or drain, efficiency. They are assumed roughly equivalent.

In the class A case, the device is biased at half the maximum voltage and half the maximum current, so the dc power dissipation is approximately $(V_{max}/2)(I_{max}/2)$. Thus, the peak power-added efficiency is

$$\eta_{PAE(max)} = \frac{P_{out(max)}}{\left(\dfrac{V_{max}}{2}\right)\left(\dfrac{I_{max}}{2}\right)} = \frac{\dfrac{V_{max} I_{max}}{8}}{\left(\dfrac{V_{max}}{2}\right)\left(\dfrac{I_{max}}{2}\right)} = \frac{1}{2} \qquad (8.20)$$

So, the absolute best power-added efficiency that can be achieved in the class A case is only 50%, and the efficiency rises linearly from zero as the output power increases.

In practice, the efficiency of the class A case will not reach that ideal level, because of the finite gain of the amplifier (increasing the RF input power) and the finite "knee" voltage of the transistor, which lowers the effective voltage swing. The effect of the knee voltage is to lower the effective maximum swing of the transistor from V_{max} to approximately $V_{max} - V_{knee}$, reducing the peak power available from the transistor and lowering its overall efficiency. Nevertheless, the class A amplifier is the most linear of all PA topologies, because the device remains "on" during the full cycle of operation; the linearity of the amplifier is then limited by the linearity of the active transistor.

In an effort to improve the efficiency of the PA, designers often explore alternative modes of operation for CDMA PAs. The class B amplifier achieves improved power-added efficiency at the cost of reduced linearity. The collector current and voltage waveforms are shown in Figure 8.18. In this case, the amplifier conducts for exactly one-half of the cycle for a conduction angle of 180 degrees. It is clear that the collector current waveforms are highly nonlinear, so filtering typically is employed between the collector and the load to eliminate the harmonic content of the output, as shown in Figure 8.19. Interestingly, the ideal class B amplifier is completely free of odd-order harmonic distortion and is therefore quite "linear" in the sense that it should generate no in-band distortion, although that condition is difficult to achieve in practice.

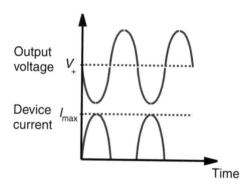

Figure 8.18 Collector current and voltage waveforms of an ideal class B amplifier.

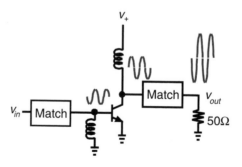

Figure 8.19 Typical class B common-emitter amplifier configuration.

The class B amplifier theoretically requires 6 dB more input RF power, because the voltage swing at the input must be doubled to achieve the same current and voltage swing. That reduction in gain is problematic for a variety of microwave PA applications, where gain is at a premium. The output of the transistor is now a rich generator of harmonics, complicating the analysis of the output considerably. In most cases, it is a good assumption that the transistor output is presented with a short circuit at all the harmonic frequencies, and an optimized load impedance only at the fundamental. That requirement is hard to achieve in practice, but it represents a good first step to further analysis.

The efficiency of the class B case can be analyzed by taking the Fourier components of the collector current waveform of Figure 8.18, that is,

$$i_C(t) = I_{max}\left\{\frac{1}{\pi} + \frac{1}{2}\sin(\omega_0 t) - \frac{2}{\pi}\left[\frac{\cos(2\omega_0 t)}{3} + \frac{\cos(4\omega_0 t)}{15} + \dots\right]\right\}$$

(8.21)

Note that the ideal class B amplifier generates only even-order distortion products. The component of the collector current at the fundamental frequency, when the device is operated at its maximum output power, is given by

$$I_{rf} = \frac{1}{2}I_{max}$$

(8.22)

The dc component of the collector current at the same maximum output power is

$$I_{dc} = \frac{1}{\pi}I_{max}$$

(8.23)

Thus, the peak power-added efficiency of the class B amplifier is

$$\eta_{PAE_{max}} = \frac{P_{rf}}{P_{dc}} = \frac{\dfrac{V_{rf}I_{rf}}{2}}{V_{dc}I_{dc}} = \frac{\dfrac{1}{2}\left(\dfrac{I_{max}}{2}\right)\left(\dfrac{V_{max}}{2}\right)}{\left(\dfrac{V_{max}}{2}\right)\left(\dfrac{I_{max}}{\pi}\right)} = \frac{\pi}{4}$$

(8.24)

which is a substantial improvement over the efficiency of the class A case.

The improvement in efficiency of the class B amplifier does not come free. The highly nonlinear collector current waveform leads to substantial distortion in the resulting output. That distortion can lead to spectral regrowth, although, as the Fourier analysis of the output current shows, such distortion consists of only even-order components in the ideal case.

An alternative implementation of the class B amplifier, which ideally is free of even-order distortion as well, is shown in Figure 8.20. In this case, two class B amplifiers are driven in a push-pull configuration, their inputs and outputs coupled together via ideally lossless transformers. Each amplifier continues to operate in the class B mode, with a peak efficiency of 78.5%, but the resulting current waveform at the load is a linear transformation of the input signal. The technique is promising, but it is rarely used in practice at microwave frequencies because the losses in the transformers tend to be excessive, reducing the intrinsic advantages of the approach.

The class A amplifier achieves acceptable linearity at the price of poor efficiency, and the class B amplifier achieves good efficiency at the price of poor linearity. A compromise is usually arrived at in the CDMA PA, in which the device is operated in class AB mode. In this case, the conduction angle is between 180 and 360 degrees, achieving acceptable linearity and improved power dissipation. The exact conduction angle and linearity usually are determined through careful experimental evaluation of the devices prior to circuit design.

Higher classes of operation, including C, D, E, F, and even S, have been proposed and are, in fact, quite common at lower frequencies [18]. However, those modes are highly nonlinear and are rarely used for linear wireless applications.

8.5.3 Devices for PAs

CDMA applications demand the highest standards of linearity from the PA. That, in turn, requires that careful attention be paid to the transistor choice for the PA, because both linearity and efficiency are required. The two most common device types for commercial CDMA applications are the GaAs metal semiconductor field effect transistor (MESFET) and the GaAs heterojunction bipolar transistor (HBT).

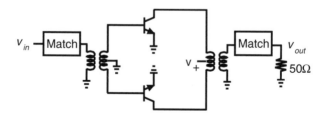

Figure 8.20 Push-pull implementation of a class B PA.

The GaAs MESFET power transistor is one of the oldest solid-state microwave PAs, with its early development dating back to the 1970s. The MESFET exhibits excellent linearity and breakdown voltage; its major drawback is a negative threshold voltage. That negative threshold voltage typically requires a separate power supply regulator to bias the gate of the transistor, as shown in Figure 8.21. That raises the cost and the complexity of the RF transmitter, a major problem with the technology. Furthermore, a P-type metal oxide semiconductor (PMOS) switch is often employed in series with the battery of a GaAs MESFET-based PA. That is because the MESFET cannot fully pinch off under most circumstances, leaving several milliamperes of drain current flowing in the standby mode of operation. That parasitic current reduces the standby time of the mobile radio and must be minimized. The series MOSFET shuts off the transistor completely, allowing for improved standby time, at the expense of some wasted power dissipated by the switch itself when the transistor is on.

The GaAs HBT is a more modern device, relying on advances in GaAs materials and device technology and exhibiting fewer of the drawbacks associated with the GaAs MESFET. In particular, like the MESFET, the HBT exhibits high gain, high linearity and outstanding breakdown voltage. However, unlike the GaAs MESFET, it does not require a negative dc bias voltage at the input (base), nor does it require the addition of a series PMOS device to ensure that the device is nonconducting during standby time. The one potential drawback of the GaAs HBT is its tendency to exhibit thermal runaway under some bias conditions [19].

That problem can be grasped intuitively by recalling that the base-emitter voltage of a bipolar transistor has a negative temperature coefficient at a constant collector current of 1–2 mV/°C. In addition, the current gain β of the device also exhibits a negative temperature coefficient. Power transistors typically consist of multiple devices in parallel, all biased from a constant current supply

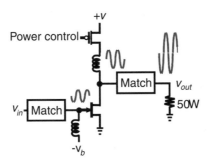

Figure 8.21 Schematic of GaAs MESFET PA illustrating bias requirements.

to the base. Because there is some natural nonuniformity in the devices, it is inevitable that one device draws slightly more base current than its neighbors and heats up. If the proportional rise in base current exceeds the drop in the current gain, the higher temperature will cause an even larger current rise, generating an even larger temperature rise, ad infinitum. The transistor reaches a point where the entire current is drawn through a single unit device, and the overall current exhibits a "collapse," as shown in Figure 8.22(a). That problem can occur simply because the devices in the middle of an array of transistors tend to be hotter than devices at the periphery.

There are several well-known solutions to the problem. One solution involves the use of cascode transistors, shown in Figure 8.22(b), which reduce the power dissipated in the collectors of the common-emitter transistors. Another approach involves the addition of "ballasting" resistors in either the emitter or the base leads, as shown in Figure 8.22(c) [20]. In that case, the voltage across the resistor rises as the current rises, reducing the base-emitter voltage and eliminating thermal runaway. The penalty is a reduction in transistor gain and power-added efficiency, although the effect usually is quite small.

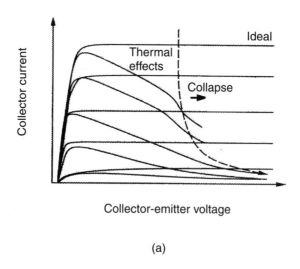

(a)

Figure 8.22 HBT power devices: (a) thermal collapse of HBT, (b) use of cascode transistors to minimize thermal runaway, (c) addition of ballasting resistors to minimize the effect of thermal runaway.

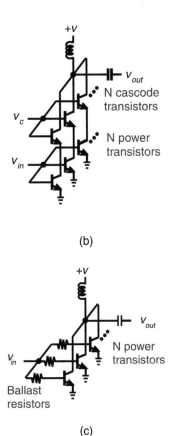

(b)

(c)

Figure 8.22 (continued).

References

[1] Couch, L., *Digital and Analog Communications Systems*, Prentice Hall.

[2] Gray, P., "Basic MOS Operational Amplifier Design," in *Analog MOS Integrated Circuits*, IEEE Press, 1980.

[3] Solomon, J., "The Monolithic Op Amp: A Tutorial Study," *IEEE J. of Solid-State Circuits*, Dec. 1974, pp. 314–332.

[4] Koullias, I. A., "A 900 MHz Transceiver Chip Set for Dual-Mode Cellular Radio Mobile Terminals," *ISSCC Digest of Tech. Papers*, Feb. 1993, pp. 140–141.

[5] Razavi, B., *RF Microelectronics*, Prentice Hall, 1998.

[6] McGee, W., "Cascade Synthesis of RC Polyphase Networks," *Proc. of 1987 IEEE ISCAS*, pp. 173–176.

[7] Crols, J., and M. Steyaert, "An Analog Integrated Polyphase Filter for a High-Performance Low-IF Receiver," *1995 Symp. on VLSI Circuit Design*, pp. 87–88.

[8] Koullias, I. A., "A 900 MHz Transceiver Chip Set for Dual-Mode Cellular Radio Mobile Terminals," *ISSCC Digest of Tech. Papers*, Feb. 1993, pp. 140–141.

[9] Rosenbaum, S., C. Baringer, and L. Larson, "Design of a High-Dynamic Range Variable Gain Amplifier for a DBS Tuner Front-End Receiver," IEEE UCSD Conf. on Wireless Communications, 1998, pp. 83–89.

[10] Schmoock, J., "An Input Stage Transconductance Reduction Technique for High-Slew Rate Operational Amplifiers," *IEEE J. of Solid-State Circuits*, Vol. SC-10, No. 6, Dec. 1975, pp. 407–411.

[11] Barrie, G., "Current-Mode Circuits From a Translinear Viewpoint: A Tutorial," in C. Toumazou, F. J. Lidgey, and D. G.. Haigh (eds.), *Analog IC Design: The Current-Mode Approach*, London: Peragrinus (on behalf of IEE), 1990.

[12] Hikita, M., et al., "High Performance SAW Filters With Several New Technologies for Cellular Radio," *Proc. Ultrasonics Symp.*, 1984, pp. 82–92.

[13] Sevic, J., "Statistical Characterization of RF Power Amplifier Efficiency for CDMA Wireless Communication Systems," *Proc. 1997 Wireless Communications Conf.*, pp. 110–113.

[14] Wu, Q., M. Testa, and R. Larkin, "Linear Power Amplifier Design for CDMA Signals," *1996 IEEE MTT Symp. Digest*, San Francisco, pp. 851–854.

[15] CDMA Development Group, "CDG Stage 4 System Performance Tests," Mar. 18, 1998.

[16] Su, D., and W. McFarland, "An IC for Linearizing RF Power Amplifiers Using Envelope Elimination and Restoration," *IEEE J. Solid-State Circuits*, Vol. 33, No. 12, Dec. 1998, pp. 2252–2258.

[17] Cristal, E. G., "Impedance Transforming Networks Of Low-Pass Filter Form," *IEEE Trans. Microwave Theory and Techniques*, MTT 13, No. 5, Sept. 1965, pp. 693–695.

[18] Kraus, H., C. Bostian, and F. Raab, *Solid-State Radio Engineering*, Wiley, 1980.

[19] Liu, W., et al., "The Collapse of Current Gain in Multi-Finger Heterojunction Bipolar Transistors: Its Substrate Dependence, Instability Criteria, and Modeling," *IEEE Trans. on Electron Devices*, Vol. 41, No. 10, Oct. 1994, pp. 1698–1707.

[20] Liu, W., et al., "The Use Of Base Ballasting to Prevent the Collapse of Current Gain in AlGaAs/GaAs Heterojunction Bipolar Transistors," *IEEE Trans. on Electron Devices*, Vol. 43, No. 2, Feb. 1996, pp. 245–251.

9

RF Receiver Circuits

The implementation of the RF receiver for a typical mobile radio represents one of the most daunting challenges in the entire transceiver design. The challenge is a consequence of the fact that the RF receiver has to accommodate a tremendous range of signal powers and to select from that range of received signals the one "desired" signal to the exclusion of all the others. The problem of selectivity is one of the classic challenges faced by designers of super heterodyne receivers.

The second challenge associated with the design of the wireless receiver is related to the sensitivity of the amplifier. This is the smallest signal that can be received with the desired SNR and hence demodulated BER. The two constraints—selectivity and sensitivity—determine the overall performance of the RF receiver.

This chapter presents the low-noise downconverter, detailing the building blocks it comprises, the LNA, and the mixer. It also describes gain control, which is needed to optimize receiver performance. Last, it covers key baseband circuits that condition the received signal for digitization and demodulation. Together, those circuits form the RF receiver shown in Figure 9.1.

9.1 RF LNAs

RF downconverter circuits typically consist of an LNA, an image reject filter, and a mixer. The design of the LNA is especially challenging, because it typically accepts a broad range of signals and frequencies from a diverse array of sources, including potentially very weak desired signals along with very large interferers. Because the LNA has to accept the widest array of signals in the receiver, it

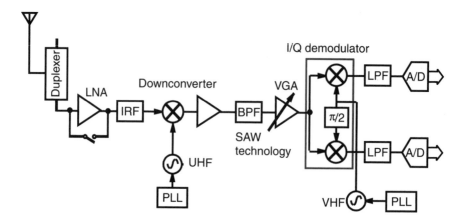

Figure 9.1 Block diagram of a typical CDMA IS95 RF receiver.

is often considered one of the bottlenecks of receiver design [1]. Table 9.1 lists the specifications for a typical mobile radio LNA.

The LNA is usually implemented with a bipolar or MESFET transistor, as shown in Figure 9.2. Other alternatives include pseudomorphic high electron mobility transistor (PHEMT) devices for extremely low-noise applications or N-type metal-oxide semiconductor (NMOS) transistors as part of a single-chip receiver. Because it is this first stage that primarily sets the Noise Figure of the entire receiver, the noise generated by the LNA must be minimized relentlessly, without compromising the linearity of the circuit. Because most linearizing feedback techniques result in added noise, the circuit approaches for realizing the LNA typically are very simple.

The Noise Figure of the LNA is a strong function of the impedance that is presented to the individual transistor. If the impedance deviates substantially from its ideal value, the Noise Figure can rise dramatically. A simple expression for that variation is given by [2]

$$F = F_{min} + \frac{g_n}{r_s}[(r_s - r_{opt})^2 + (x_s - x_{opt})^2] \tag{9.1}$$

Table 9.1
Specifications for a Typical CDMA IS95 Receiver LNA

Gain	>16 dB
Noise Figure	<2.5 dB
Input/output VSWR	<2:1
Input intercept point (dBm)	0 dBm
Reverse isolation	25 dB
Frequency	850 MHz, 1,900 MHz

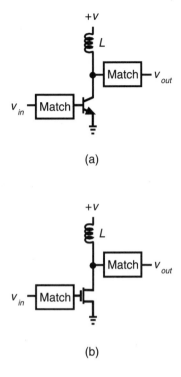

(a)

(b)

Figure 9.2 Implementation of LNAs using (a) bipolar or (b) MESFET technologies.

where F_{min} is the minimum value of Noise Factor[1] when the device is presented with its optimum source impedance, r_s is the real part of the source impedance, whose optimum value is r_{opt}, and x_s is the imaginary part of the source impedance, whose optimum value is x_{opt}. The quantity g_n has units of conductance and is a device-specific parameter that determines how quickly the Noise Figure rises from its minimum when the source impedance varies. Typical microwave transistors are characterized by values of F_{min}, g_n, r_{opt}, and x_{opt}. Given those parameters, the Noise Figure of an LNA can be calculated for any arbitrary source impedance value.

Several different device types can be employed for the design of the LNA. A silicon bipolar junction transistor (BJT) or an HBT is often used for LNA applications. A cross-section of the device and its equivalent circuit representation are shown in Figure 9.3.

The minimum Noise Figure for a BJT in the common-emitter mode is given by [3]

---

1. Noise Figure and Noise Factor are related by the expression, $NF = 10 \log F$.

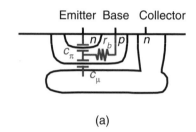

(a)

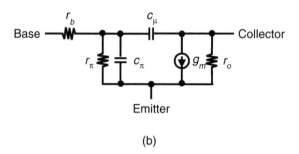

Emitter

(b)

Figure 9.3 (a) Cross-section of silicon BJT and (b) its high-frequency equivalent circuit.

$$F_{min} = 1 + \frac{(1 + g_m r_b)}{\beta_0} + \frac{\omega^2 r_b C_\pi^2}{g_m} + \sqrt{\frac{(1 + g_m r_b)}{\beta_0} + \frac{2\omega^2 r_b C_\pi^2}{g_m} + \frac{\omega^4 r_b^2 C_\pi^4}{g_m^2}}$$

$$(9.2)$$

where g_m is the device transconductance (typically I_c/V_T), r_b is the base resistance, β_0 is the low-frequency current gain of the device, ω is the operating frequency in radians (where $\omega = 2\pi f$), and C_π is the base-emitter capacitance. The ratio of C_π to g_m is approximately the forward transit time (τ_F), and the reciprocal of that quantity is related to the unity current gain frequency ($\omega_T = 1/\tau_F$). Equation (9.2) suggests that the route to achieving a low-noise figure involves the use of a transistor with a high ω_T and low base resistance. At the same time, the Noise Figure of the device inevitably will increase with frequency. Typical values for f_T for a modern BJT are in the range of 25–50 GHz.[2]

Several interesting conclusions can be drawn from that result. One is that, at sufficiently low frequencies, the minimum Noise Factor of the BJT reduces to

---

2. The unity gain frequency is expressed in hertz, not radians per second, with $f_T = \omega_T/2\pi$.

$$F_{min} = 1 + \frac{(1 + g_m r_b)}{\beta_0} + \sqrt{\frac{(1 + g_m r_b)}{\beta_0}} \qquad (9.3)$$

Thus, the goal of high dc current gain (β_0) and low base resistance is clear for the minimization of device Noise Figure.

Equation (9.2) also can be employed to determine the optimum transconductance—and hence the dc current—of the device, which is achieved when the Noise Factor is at a minimum and is given by

$$g_{m(opt)} = \frac{\omega C_{je}\sqrt{\beta_0}}{\sqrt{1 + (\omega\tau_f)^2\beta_0}} \qquad (9.4)$$

where C_{je} is the depletion capacitance at the emitter-base junction, and τ_f is the electron transit time in the base. The transconductance of the device can be altered by varying the dc bias current ($g_m = I_c/V_T$). These equations demonstrate that, for a BJT, the optimum noise figure is obtained from a device exhibiting low base resistance and high ω_T. Finally, the optimum source resistance and reactance (the impedances presented to the base of the device) are given by

$$r_{opt} = \frac{\sqrt{g_m 2((1 + 2g_m r_b)/\beta_0) + r_b\omega^2 C_\pi^2(2g_m + r_b\omega^2 C_\pi^2)}}{\dfrac{g_m^2}{\beta_0} + \omega^2 C_\pi^2} \qquad (9.5)$$

and

$$x_{opt} = \frac{\omega C_\pi}{\dfrac{g_m^2}{\beta_0} + \omega^2 C_\pi^2} \qquad (9.6)$$

The optimized design of the LNA then involves presenting the required impedance to the device to achieve the minimum Noise Figure. Because x_{sopt} typically is positive (inductive), the impedance match usually is accomplished by an inductor in series with the base, as shown in Figure 9.4. Fortunately, the value of reactance has the sign opposite that of the input impedance of the transistor, so the optimum imaginary portion of the source impedance minimizes Noise Figure and maximizes power transfer into the device. Note, however, that at high frequencies, the magnitude of the optimum value x_S

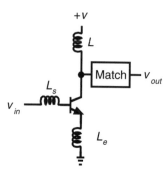

Figure 9.4 Illustration of a series inductor in the base and the use of inductive series feedback in the emitter to improve the impedance match of an LNA.

drops with frequency, whereas an inductor will tend to increase its value of reactance (x) with frequency, which makes it difficult to achieve an optimum broad bandwidth noise impedance match.

Note that the real portion of the optimum source impedance of the BJT does not approach 50Ω except in very unusual circumstances. That will result in some mismatch loss of the available signal power to the input of the LNA, as well as a relatively high input VSWR. However, inductive series feedback in the emitter can also be added to the device to improve the resulting power transfer *and* linearity, at the expense of somewhat lower gain, as shown in Figure 9.4.

In that case, the F_{min} of the device is only slightly altered; in fact, it may be slightly lower than the original circuit, since the inductor adds no noise of its own. However, the input impedance of the device is raised to

$$Z_{in} \approx r_b + \omega_T L_e + \frac{1}{j\omega C_\pi} \qquad (9.7)$$

Now, the real portion of the transistor input impedance can be set to r_{opt} through judicious choice of L_e. The required real portion of the optimum source impedance r_{opt} remains approximately unchanged by that feedback, and the imaginary portion of the impedance is raised slightly by the additional impedance. As a result, the real portion of the source impedance required by the device to achieve minimum noise figure and the input impedance of the device are now the same, and an appropriate impedance transformation at the input will result in a nearly ideal power match to the transistor.

Similar calculations can be employed to calculate the minimum noise figure and optimum source impedance of a CMOS device. A simplified cross-section and schematic of the equivalent circuit and noise model of a CMOS transistor are shown in Figure 9.5, which illustrates the major source of noise in the device.

In this case, the minimum Noise Figure is given by [4]

$$F_{min} \approx 1 + \frac{2}{\sqrt{5}} \frac{\omega}{\omega_T} \sqrt{\gamma \delta (1 - |c|^2)} \tag{9.8}$$

where γ is the "excess" Noise Factor of the drain-source noise current and has a value of 2/3 for long-channel devices, rising dramatically for shorter channel devices to values greater than unity [5]. The quantity δ accounts for the channel-induced gate noise that appears in a MOSFET due to the capacitive coupling between the gate and the channel. This quantity has a value of about 4/3 (for long-channel devices) and rises as the gate length of the devices is reduced. There is some evidence that the ratio between γ and δ remains at approximately 2 as the channel length of the devices scale [6]. Finally, because the gate and drain noise currents are partially correlated, the quantity c is the correlation coefficient between the gate and drain noise, defined as

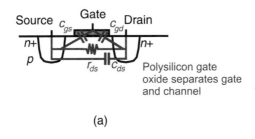

(a)

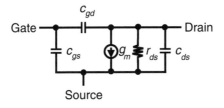

(b)

Figure 9.5 (a) Simplified cross-section and (b) small-signal model of MOSFET, showing sources of noise.

$$c \equiv \frac{\overline{i_{ng} i_{nd}^{*}}}{\sqrt{|i_{ng}^2|} \sqrt{|i_{nd}^2|}} \qquad (9.9)$$

which is approximately 0.4j for long-channel devices. The expression for F_{min} illustrates the importance of a high device f_T to achieve a low device noise figure, which is in good agreement with the result obtained in the case of a BJT. Equation (9.8) does not include the sources of noise associated with the ohmic contact resistances to the intrinsic device, that is, the gate, drain, and source resistances. Those can be added to the model in a straightforward manner and are particularly important for operation in the short-channel regime [7].

At the same time, the optimum source conductance and susceptance of the MOSFET can be given by

$$g_{opt} = \alpha \omega C_{gs} \sqrt{\frac{\delta}{5\gamma}(1 - |c|^2)} \qquad (9.10)$$

and

$$b_{opt} = -\omega C_{gs} \left(1 + \alpha |c| \sqrt{\frac{\delta}{5\gamma}} \right) \qquad (9.11)$$

where α is the ratio of the device transconductance (g_m) to the device zero drain bias drain-source conductance (g_{ds0}), which is approximately unity for long-channel devices. As was the case with the BJT, the optimum source susceptance is approximately the conjugate of the transistor input susceptance, providing for a nearly optimum imaginary impedance match. However, the real part of the input admittance is nowhere near a conjugate match, and the use of inductive feedback is often required if a low input VSWR is desired. In fact, the real part of the input admittance of a MOSFET is typically much higher than that of a BJT, raising the difficulty associated with the impedance match. Typically, the same impedance matching techniques used for the BJT work well for the MOSFET.

The Noise Figure of a GaAs MESFET is more difficult to determine analytically, based on first operating principles of the device. That is due to the short-channel ($<0.25\mu$m) operation of a typical low-noise GaAs MESFET, as well as the importance of the extrinsic elements in the device operation. Figure 9.6 shows a cross-section of a typical GaAs MESFET and a typical equivalent circuit of the GaAs MESFET. Several noise models have been developed for the GaAs MESFET in recent years; the approach presented by

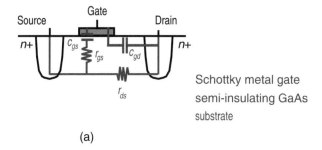

Schottky metal gate
semi-insulating GaAs
substrate

(a)

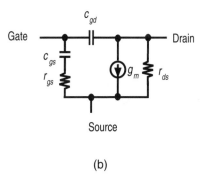

(b)

Figure 9.6 (a) Simplified cross-section and (b) small-signal equivalent circuit model of a GaAs MESFET, showing sources of noise.

Hughes [8] provides an excellent fit to a wide range of devices. In that case, the two sources of noise in the device (r_{gs} and r_{ds}) have a unique noise temperature[3] associated with them, and the noise current of each element is obtained through the typical expression $i_n^2 = 4kT_{eq}/R$.

The noise temperature associated with the input circuit (T_{gs}) is the ambient temperature of the device, and the noise temperature associated with the output circuit (T_{ds}) is at a higher temperature, associated with the high-energy nonequilibrium transport of electrons through the channel. Typical values of T_{ds} range from 250°C to 600°C. In this case, the minimum Noise Figure is given by the very simple expression

$$F_{min} = 1 + \frac{\sqrt{T_{gs} T_{ds}}}{T_0} \frac{\omega}{\omega_{max}} \qquad (9.12)$$

3. Noise temperature is another way to express excess circuit noise and Noise Factor (see Section 7.1.4).

where ω_{max} is the frequency at which the maximum power gain of the device is unity and is given by

$$\omega_{max} \approx \frac{\omega_T}{\sqrt{4r_{gs}g_{ds}}} \qquad (9.13)$$

As in the case of the other two devices, the optimum source reactance is a complex conjugate of the device input reactance. The real portion of the optimum source impedance is given by

$$r_{opt} = r_{gs}\sqrt{\frac{4T_{gs}}{T_{ds}}}\frac{\omega_{max}}{\omega} \qquad (9.14)$$

Historically, the GaAs MESFET exhibits a lower Noise Figure than a bipolar device, due to its lower gate resistance (compared to the base resistance of a bipolar device) and the absence of shot noise in the drain or gate region. It also exhibits a better Noise Figure than a silicon MOSFET because of its higher ω_T and lower gate resistance. That lower Noise Figure is achieved along with a somewhat higher cost of production, rendering the GaAs MESFET most suitable for implementation in hybrid or small-scale integrated circuit form.

In most applications, the linearity of the LNA, as measured by its IP_3, is at least as important as its Noise Figure. The linearity of the circuit is difficult to predict analytically and usually is obtained through simulation. However, some general conclusions about the linearity behavior of transistor amplifiers can be obtained, although their range of applicability must be carefully verified. Section 7.1.5 presented a simple model of the nonlinearity of an amplifier and analyzed the resulting intermodulation performance through a power series approximation. In this case, the transfer function of the amplifier is

$$v_o = a_1 v_i + a_2 v_i^2 + a_3 v_i^3 + \dots \qquad (9.15)$$

and the IIP_3 is given by

$$V_{IP} = \sqrt{\frac{4}{3}\frac{a_3}{a_1}} \qquad (9.16)$$

where a_1 is the first-order coefficient of the power-series expansion of the amplifier gain, a_2 is the second-order coefficient of the power-series expansion of the amplifier gain, and a_3 is the third-order coefficient of the power-series expansion of the amplifier gain.

We can model the effect of feedback on the linearity of this circuit in a straightforward manner, as shown in Figure 9.7, where a linear feedback term subtracts a portion of the output from the input. In this case, the resulting input signal to the amplifier is given by

$$v_i = s_{in} - fv_o \tag{9.17}$$

where f is the feedback network transfer function. In the case of a common-emitter or common-source transistor, the feedback factor f is the impedance of the network, and the forward gain is the nonlinear transconductance of the device (g_m).

The output transfer function of the final amplifier can be given by

$$v_o = b_1 v_i + b_2 v_i^2 + b_3 v_i^3 + \dots \tag{9.18}$$

where

$$b_1 = \frac{a_1}{1 + a_1 f} \tag{9.19}$$

and

$$b_3 = \frac{a_3(1 + a_1 f) - 2a_2^2 f}{(1 + a_1 f)^5} \tag{9.20}$$

That reveals some interesting features about the nonlinear behavior of the feedback amplifier. Even if $a_3 = 0$, b_3 is finite. In other words, the addition of feedback can create third-order distortion even if the original amplifier had none. That can happen in CMOS amplifiers, which intrinsically have very low third-order distortion. On the other hand, b_3 can also be set to zero, completely eliminating third-order distortion. That occurs when $a_3(1 + a_1 f) = 2a_2^2 f$.

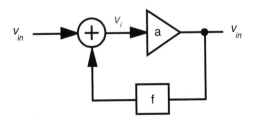

Figure 9.7 Model of linear feedback for a nonlinear amplifier.

Some circuits attempt to achieve this condition, but in practice it is extremely difficult to maintain it over process and temperature variations.

If $a_2 = 0$, the relative level of IMD3 and desired output is

$$IMD_3 = \frac{3}{4}\frac{a_3}{a_1}\frac{1}{(1 + a_1 f)^3}V_{in}^2 \qquad (9.21)$$

so the IMD_3 of the feedback amplifier is reduced by $(1 + a_1 f)^3$—a substantial benefit.

In the case of a common-emitter or common-source amplifier with series feedback, the feedback factor is simply the impedance of the feedback element, either resistive or inductive, or some combination of the two. That improvement in linearity generally applies whether the feedback is inductive or resistive. If the feedback is resistive, the Noise Figure is degraded by the addition of the thermal noise due to the resistor, and the resulting tradeoff between noise and linearity is straightforward. If the feedback is inductive, the feedback does not add any noise of its own, yet the linearity is improved. In the case of a BJT, it has been demonstrated that the minimum distortion occurs at a frequency given by [9]

$$f = \frac{1}{\sqrt{2C_{je}(L_b + L_e)}} \qquad (9.22)$$

which is also approximately the frequency at which the minimum Noise Figure occurs. It has also been demonstrated that the simultaneous achievement of low-noise figure and low distortion is a substantial advantage of the common-emitter configuration compared with the common-base configuration [9]. Similar arguments apply for the CMOS and GaAs MESFET configurations, which typically exhibit the best response in the common-source configuration. The exact expression for the distortion of an inductively degenerated BJT amplifier is extremely involved but can be found in [10].

9.2 Downconversion Mixers

The design of the downconversion mixer is complicated by a number of factors and, despite its seemingly simple function, requires some fairly sophisticated analysis. The most important aspect of the mixer operation is the translation of a high-frequency carrier (at RF) to a low-frequency carrier (at IF). That relationship is shown in Figure 9.8.

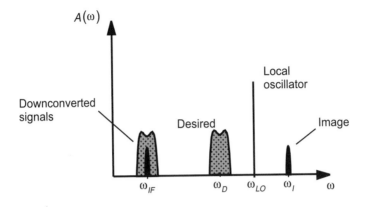

Figure 9.8 Frequency domain illustration of downconversion mixer operation.

In the simplest implementation, the mixing function can be viewed as an ideal multiplier, whose output is given by

$$v_{out}(t) = v_{in}(t)v_{to}(t) \tag{9.23}$$

In practice, however, the use of an ideal multiplier for the downconversion operation has a number of drawbacks, especially the resulting noise, which is very high for fundamental reasons. Therefore, a higher performance model of the downconversion process is the doubly balanced modulator, as shown in Figure 9.9. In that case, the output of the amplifier is periodically connected to either the +1 or −1 gain stage at the LO frequency. The output is a replica of the input, multiplied by ±1 at the LO frequency, that is,

$$v_{out}(t) = [v_{in+}(t) - v_{in-}(t)]\frac{4}{\pi}\sum_{n=1,3,5,\ldots}^{\infty}\frac{1}{n}\cos(n\omega_0 t) \tag{9.24}$$

Notice that in this ideal case, the output of the mixer at the LO frequency is completely suppressed if the input has no dc component (which is typically

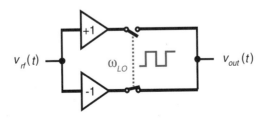

Figure 9.9 Doubly balanced modulator employed as a downconversion mixer.

the case), so the LO-IF isolation is especially good. In addition, the double-balanced modulator intrinsically suppresses all even harmonics of the LO and can suppress all even harmonics of the input signal as well. Alternatively, the input signal can be multiplied by 1 and 0, as in the single-balanced modulator design, whose output is given by

$$v_{out}(t) = [v_{in+}(t) - v_{in-}(t)] \left[\frac{1}{2} + \frac{2}{\pi} \sum_{n=1,2,3,\ldots}^{\infty} \frac{(-1)^n}{n} \cos(n\omega_0 t) \right]$$

$$(9.25)$$

That singly balanced version of the modulator will be sensitive to even harmonics of the LO but insensitive to even harmonics of the input signal. Alternatively, another version of the single-balanced version of the modulator could have the transfer function

$$v_{out}(t) = [V_{DC} + v_{in}(t)] \frac{4}{\pi} \sum_{n=1,3,\ldots}^{\infty} \frac{1}{n} \cos(n\omega_0 t) \qquad (9.26)$$

where the modulator has responses only to odd harmonics of the LO but has responses to all harmonics of the input signal.

Finally, an unbalanced modulator or mixer will have a response of the form

$$v_{out}(t) = [V_{DC} + v_{in}(t)] \left[\frac{1}{2} + \frac{2}{\pi} \sum_{n=1,2,3,\ldots}^{\infty} \frac{(-1)^n}{n} \cos(n\omega_0 t) \right] \quad (9.27)$$

where the output is sensitive to all the harmonics of both the RF input and LO waveforms.

The discussion of noise figure, confusing in the best of circumstances, can take an especially bizarre turn in the case of mixers. That is due to the original definition of Noise Figure (from Section 7.1.4) as

$$F = \frac{(S/N)_{in}}{(S/N)_{out}} \qquad (9.28)$$

The complication occurs because the frequency spectrum of the desired signal can be above, below, or centered at the frequency of the LO. In the case where the desired signal is above or below the LO, the mixer converts the signal plus noise at that frequency to the IF output. It also converts any noise at image frequency to the same IF output frequency. As such, the noise contribution is twice that which is expected (or 3 dB higher). That measurement is referred to as single-sideband (SSB) noise figure. In the case in which the desired signal is centered about the frequency of the LO, there is no image signal and the noise figure measurement (known as double sideband, or DSB) is straightforward. The excess noise figure of 3 dB is the source of endless confusion in mixer measurements and characterization, because even an ideal noiseless mixer will exhibit a finite 3-dB SSB noise figure. Figure 9.10 illustrates these noise figure measurement issues.

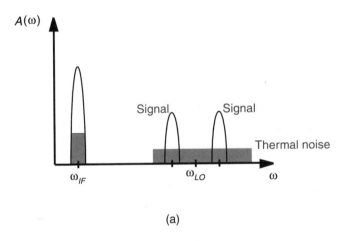

(a)

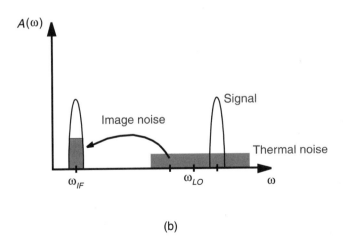

(b)

Figure 9.10 Illustration of noise downconversion process in mixers and the resulting noise figure: (a) DSB measurement and (b) SSB measurement.

Mixers are classified as either passive or active. Each type offers different advantages and disadvantages, which are outlined next.

9.2.1 Passive Mixer Design

The classic single-balanced and double-balanced diode switching mixers, shown in Figure 9.11, implement the balanced modulator through a switching operation. This mixer is a passive circuit—the diodes provide no amplification of the signal—so the output of the mixer can closely approximate the results of (9.24) and (9.25). The linearity of the diode-based single-balanced and double-balanced mixer is outstanding and depends primarily on the power level of the LO signal, as well as the cutoff frequency and series resistance of the diodes. The major source of the nonlinearity in the circuit is the variable resistance of the forward-biased diodes, which is minimized through a high forward bias current and hence a high LO power. Typical power levels for the LO are between +5 and +20 dBm, and typical input intercept points are also between

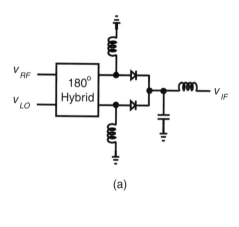

(a)

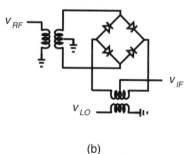

(b)

Figure 9.11 Schottky diode mixers: (a) single-balanced and (b) double-balanced configurations.

+5 and +20 dBm; the input intercept point and the LO power track each other closely. Improvements in input intercept point performance can be achieved by placing additional diodes in series, although at the expense of a higher required input power level. Because the mixers are inherently passive devices, their DSB noise figures are very close to the mixer loss, which also decreases with increasing LO power.

Figure 9.12 is a cross-sectional diagram and an equivalent circuit model for a typical Schottky diode. The current through the diode is typically assumed to be

$$I(V) = I_s(e^{(V/\eta V_T)} - 1) \qquad (9.29)$$

where V_T is the thermal voltage (approximately 26 mV at room temperature) and η is the diode ideality factor, approximately 1. The junction capacitance is usually approximated by the expression

$$C(V) = \frac{C_{j0}}{1 - \left(\dfrac{V}{V_{bi}}\right)^{\gamma}} \qquad (9.30)$$

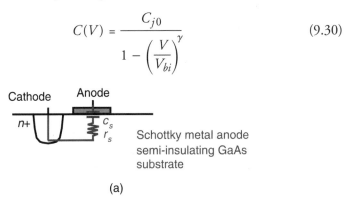

(a)

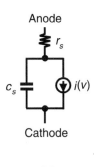

(b)

Figure 9.12 (a) Simplified cross-sectional view and (b) equivalent circuit of Schottky diode.

where C_{j0} is the junction capacitance, V_{bi} is the built-in potential of the Schottky diode, and γ is related to the doping gradient of the semiconductor material and is typically between 0.3 and 0.7. In this case, it is desirable to minimize both the zero-bias junction capacitance (C_{j0}) and the series resistance (r_s). Unfortunately, minimizing the capacitance by reducing the diode area also increases the series resistance, so improvements have to come in the vertical design of the device, either through improved materials or improved design of the epitaxial layer. The loss in the diode due to its series resistance can be approximated by [11]

$$\delta = 1 + \frac{r_s}{z_s} + \frac{z_s f_{RF}^2}{r_s f_c^2} \qquad (9.31)$$

where z_s is the source impedance, f_{RF} is the frequency of operation, and f_c is the cutoff frequency of the Schottky diode. Typical cutoff frequencies for microwave Schottky diodes are in the range of 100 to 1,000 GHz.

It has been pointed out [12] that there is an optimum value of r_s that minimizes δ, which occurs when

$$r_s = z_s \frac{f_{RF}}{f_c} \qquad (9.32)$$

and the optimum value of r_s is usually no more than a few ohms.

The proper impedance termination of the passive diode mixer is key to obtaining the best possible performance from the device. In this case, the reflection due to the load impedance will get mixed and delivered to the input in the same manner that the desired input is mixed and delivered to the output. A set of new frequencies is created by the mixing products, which then experience further reflection and re-reflection, ad infinitum, as illustrated in Figure 9.13. The only way to eliminate that situation is to present a 50Ω termination to the device at all frequencies—a clear impossibility. Instead, careful choice of the terminations usually can minimize the problem to an acceptable level.

Figure 9.14 shows an FET version of the classic double-balanced diode mixer. The diodes have been replaced by passive series MOSFETs, and the LO drives the gates of the transistors, alternatively turning them on and off [13]. In this case, the linearity of the mixer is outstanding, but it is very difficult to achieve comparably low series resistances to a high-frequency Schottky diode with a silicon MOSFET. As a result, the noise figure and loss of the resulting structure are high, although the linearity is very good—the typical IIP$_3$ is in excess of 0 dBm.

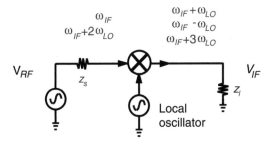

Figure 9.13 Illustration of the impedance termination challenge in Schottky mixers. Careful attention must be paid to proper termination at all the mixing products of the system.

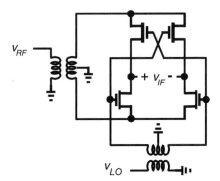

Figure 9.14 Double-balanced MOSFET mixer.

Alternative implementations of the passive FET-based mixer involve a single GaAs FET, operated in the resistive mode, in which the LO drives the gate of the common-source device, and the mixing occurs due to the time-varying resistance of the channel [14]. An example of this approach is shown in Figure 9.15. This technique achieves extremely good linearity and noise performance, although it requires outstanding performance from the FET, which is why GaAs FETs typically are employed.

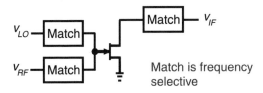

Figure 9.15 A passive FET mixer, which uses the time-varying channel resistance to achieve mixing operation.

A pair of single FET mixers, suitably combined with an 180-degree hybrid, can be used to realize a single-balanced mixer, as shown in Figure 9.16. The drawback of this approach is that the IF output, which is typically at a relatively low frequency, requires its own hybrid circuit, unlike the diode mixer case [15]. The balanced operation has the advantage of insensitivity to even harmonics, at the expense of higher LO power required for a given linearity and noise figure.

9.2.2 Active Mixer Design

The unbalanced passive FET mixer in Figure 9.15 can be employed in the active mode of operation, simply by increasing the drain voltage, to the point where the transistor is in its normal bias range [16]. The mixing operation is accomplished by the time-varying transconductance rather than the time-varying resistance, as in the passive FET case. Operation of the FET mixer in this mode tends to have worse linearity than the passive FET case, although the mixer does exhibit gain, which can be beneficial in some cases. The impedance matching of an active FET mixer must be carefully optimized, with special attention paid to the impedances at the LO and RF frequencies [17].

The unbalanced FET mixer in Figure 9.15 has a disadvantage in that the isolation between the LO and RF ports is intrinsically poor, because both signals are applied to the same terminal (the gate). This drawback is circumvented by the use of a dual-gate GaAs MESFET, as shown in Figure 9.17. In that case, the LO is applied to one gate terminal, and the RF input

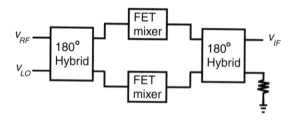

Figure 9.16 A pair of single FET mixers combined to realize a single-balanced FET mixer.

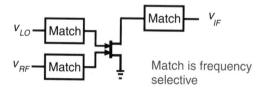

Figure 9.17 An unbalanced dual-gate GaAs FET mixer.

is applied to the other gate terminal. The performance of the dual-gate mixer is extremely complicated, and the circuits have proved difficult to optimize [18].

An alternative mixer implementation, more suitable for integrated circuit implementation, is a transistor-based design. Such designs, which can be either singly balanced or doubly balanced, are shown in Figure 9.18. The linearity and noise figure of the transistor-based mixer are determined mostly by the input devices. The input-referred linearity of the singly balanced case depends on the input intercept point performance of the input device. In the case of a bipolar mixer, the linearity of the input stage is determined by some of the same issues that affected the linearity of the LNA. Resistive or inductive feedback, as shown in Figure 9.18(c), can be employed to improve the linearity, at the expense of increased noise in the case of resistive feedback.

One technique in particular has been developed recently for the improvement of linearity in the doubly balanced mixer: the multitanh approach [19], shown in Figure 9.19. A single differential pair has a transconductance response or gain curve that falls off quickly outside a narrow range of input voltages. Adding parallel differential pairs, with different offsets, can create an aggregate transconductance response that is roughly constant over a wider range of input voltages.

A classical bipolar differential pair amplifier produces a differential output current that follows a tanh response to input voltage. To offset the tanh response and the peak of the transconductance curve, the emitter areas of the transistors that form the differential pair are sized differently. The offset in the tanh response is $V_T \ln A$, where A is the ratio of transistor emitter areas. The multitanh approach is also applicable to CMOS differential pairs. In these circuits, the offset is formed by the W/L ratio of the devices.

The small-signal transconductance of a two-stage multitanh circuit, or doublet, is given by

$$G_m = \frac{I_T}{2V_T} \frac{2A}{(1 + A)^2} \qquad (9.33)$$

and is controlled by I_T, the tail current. This circuit has increased dynamic range compared to a simple differential pair.

The linearity of the Gilbert mixer is relatively poor compared to passive mixer approaches and is essentially limited by the same constraints of the common-source or common-emitter LNA. In those cases, there is an unpleasant tradeoff between dc power consumption and linearity. This is especially difficult in a wireless communications system, because the input intercept point of the mixer must be larger than the product of the input intercept point of

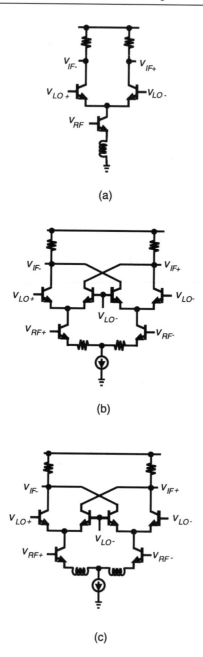

(a)

(b)

(c)

Figure 9.18 Schematic of active bipolar mixers, suitable for integrated circuit
implementation, (a) single-balanced design, (b) double-balanced (Gilbert)
configuration, and (c) Gilbert configuration with emitter feedback to improve
linearity and input isolation.

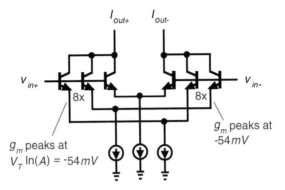

Figure 9.19 Schematic diagram of multitanh differential pair, illustrating improvement in linearity.

the preceding LNA and the gain of the LNA, to avoid degradation of the overall linearity performance.[4] On the positive side, the Gilbert mixer typically does not require the same level of LO power as a passive mixer, so the dc power required to produce the high LO power of the diode mixer is eliminated.

Many practical issues are required to make the RF performance of the Gilbert mixer adequate for most demanding applications [20, 21]. In particular, the size of the upper switching devices can have a significant impact on the linearity and noise figure of the resulting circuit. In the case of a bipolar implementation, the emitter area of the switching devices should be minimized to reduce the junction capacitance and speed the switching behavior. On the other hand, the base resistance of the switching devices should be minimized to minimize their noise contributions, which in turn implies that the device size should be increased. The optimum sizing of these devices is best obtained through careful simulation of the circuit.

9.3 Automatic Level Control

The daunting challenge in mobile radio design is the wireless communications environment. Various users, of different signal powers, unintentionally clutter the communication channel and wreak havoc on the RF receiver. In addition, the received signal strength of the desired signal varies rapidly and in an unpredictable fashion.

The RF receiver must cope with the changes in the desired signal level and also changes in any interfering signal levels. The AGC loop (see Section

---

4. Cascaded IP$_3$ is detailed in Section 7.1.5.

5.2.3) serves that purpose and relies on a VGA in the RF receiver, with more than an 85-dB gain adjustment.

The VGAs in the RF receiver differ slightly from those found in the RF transmitter. That is due primarily to system requirements. In the transmitter, output linearity was crucial, whereas in the receiver, input linearity is critical. That is because the amplifiers in the receiver must be capable of handling strong interfering signals without distortion, often providing attenuation instead of gain and placing a larger burden on input linearity than on output linearity.

Figure 9.20(a) shows one approach to VGAs in the RF receiver. The circuit is a multitanh amplifier, which offers extended input range. The bias currents, I_1 and I_2, control the gain of the circuit. Furthermore, as mentioned previously, paralleling additional differential pair amplifiers, offset from one another, can further extend the input range.

A second approach is shown in Figure 9.20(b). In that circuit, the metal oxide semiconductor (MOS) transistor M_1 simulates a variable resistor, providing emitter degeneration (local feedback) to the differential pair amplifier and thus directly increasing the linearity of the amplifier. The channel resistance of the MOS transistor is set using a replica device M_2. Both the in-circuit device and the replica transistor share the gate connection, which is driven by an operational amplifier as part of a servo loop. The servo loop equalizes the voltage drops across a known resistor R_1 and the replica transistor. As a result, $I_1 R_1 = I_2 rds_2$ and

$$r_{ds1} = \left(\frac{I_1}{I_2}\right) R_1 \qquad (9.34)$$

where I_1, I_2, and R_1 are defined in the circuit. The local feedback is adjusted by the ratio of variable currents I_1 and I_2.

In practice, a second gain control point is needed in the RF receiver to reduce front-end gain. In some situations, interfering signals can be strong enough to drive the receiver into compression before the VGA. To avoid that, a switch—controlled by the AGC algorithm—is added to bypass the LNA, as shown in Figure 9.21. The switch is used in high-signal conditions and generally is implemented using FET technology.

9.4 I/Q Demodulator

Modern modulation schemes, such as BPSK, QPSK, and GMSK, use the phase of the carrier to convey information. If the received signal is separated into orthogonal components, phase detection of the input signal is straightforward.

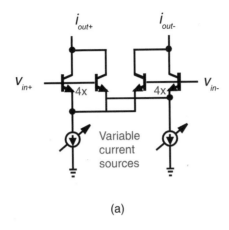

(a)

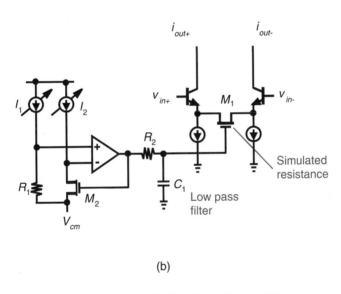

(b)

Figure 9.20 VGAs for the RF receiver: (a) multitanh amplifier and (b) variable-degeneration amplifier.

An I/Q demodulator downconverts the IF signal and splits the baseband waveform into its I and Q components. The resulting signal is then converted to digital form. The schematic of the receiver, including the I/Q demodulator, was shown in Figure 9.1. The baseband portion of the receiver typically consists of an I/Q demodulator, an analog filter, and an A/D converter.

The I/Q demodulator can be implemented using analog or digital techniques. Analog methods are subject to impairments that produce two basic

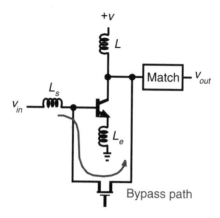

Figure 9.21 Switched-gain LNA for handling high-level signals.

effects: increased receiver interference via adjacent-channel leakage [22] and increased BER. A completely digital method requires a high-performance analog-to-digital converter, with IF sampling at four times the data rate, and wide dynamic range. In practice, CDMA IS95 radio receivers use analog I/Q demodulation.

Impairments to the I/Q demodulator can create an interfering image signal, where the energy from the quadrature channel can "leak" into the I channel and vice versa. That effect, referred to as adjacent-channel leakage, produces residual sideband energy equal to

$$RSB = \frac{1 - 2\sqrt{\Delta A/A}\cos\Delta\theta + \Delta A/A}{1 + 2\sqrt{\Delta A/A}\cos\Delta\theta + \Delta A/A} \qquad (9.35)$$

where $\Delta A/A$ is the power gain ratio and $\Delta\theta$ is the phase mismatch. Note that (9.35) matches (8.3), which describes leakage in the I/Q modulator. In practice, the residual sideband energy typically is 30 to 35 dB below the desired spectrum [23], which is acceptable for most applications.

The design of the I/Q demodulator parallels the approach taken for the I/Q modulator described in Section 8.1.2.

9.5 Baseband Channel Select Filters

The desired channel is selected by the receiver using an IF SAW filter and an integrated baseband filter. The design of the baseband filter is based on standard techniques [24–26] and specific CDMA IS95 issues.

In any filter design, the shape factor and group delay characteristics are important considerations. The characteristics generally are mapped to one of four filter prototypes (listed in Table 9.2) that optimizes performance in some aspect. In phase-modulated systems, phase linearity is crucial, while in high-interference environments, stop-band rejection is important.

The baseband filters can be positioned before or after the A/D converters; the decision affects the dynamic range requirements on the A/D converters [27]. Analog filters attenuate interfering signals and thus lower A/D converter requirements. Without the filters, the A/D converters need to transform the desired signal plus any interfering signals to the digital domain, where digital filters isolate the desired signal. In practice, the interfering signals can be 35 dB higher than the desired signal,[5] which translates to more demanding A/D converter requirements. As such, analog filtering typically is used, although advances in $\Delta\Sigma$ modulator A/D converters are shifting filtering to the digital domain.

Analog filters find use in a variety of applications, including PLLs, A/D converters, and D/A converters. The filters provide either discrete-time or continuous-time operation, although discrete-time filters typically are not used in wireless communications due to clock feedthrough, high substrate noise, and increased cross-talk and interference [28–30].

Two continuous-time filters are commonly used for wireless communications: active RC and transconductance C. Active-RC filters are traditional filter structures consisting of resistors, capacitors, and active gain stages. In these filters, the gain stages typically are operational amplifiers and tuning steps are discrete. A common active-RC filter is the Sallen-Key topology, shown in Figure 9.22(a). It finds widespread use because it has a unity gain and therefore

Table 9.2
Comparison of Filter Prototypes

Prototype	Magnitude Response	Selectivity	Phase Linearity
Butterworth	Maximally flat	Moderate	Acceptable
Chebyshev	Equal passband ripple	Maximum for all-pole structure	Poor
Bessel	Flat	Poor	Excellent
Elliptic	Equal ripple	Maximum	Poor

---

5. An interfering signal can be as much as 70 dB higher than the desired signal at the antenna but generally is reduced 35 to 40 dB by the IF SAW filter.

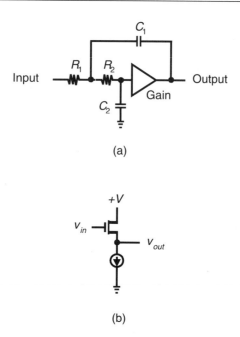

(a)

(b)

Figure 9.22 Common active-RC filter: (a) Sallen-Key structure and (b) gain stage implemented by simple follower stage.

can be implemented by a simple emitter follower or source follower, as shown in Figure 9.22(b).

Transconductance-C filters are the most popular integrated continuous-time filter structure. They use an integrator involving a transconductor and a capacitor as basic building blocks to simulate inductance. The concept is developed below. The following fundamental equations describe an inductor and a capacitor:

$$v_l = L\frac{di_l}{dt} \qquad i_c = C\frac{dv_c}{dt} \tag{9.36}$$

Notice that the two expressions in (9.36) are similar when voltage and current are interchanged. As a result, a circuit that interchanges the variables enables a capacitor to simulate an inductor. The interchange is possible in a circuit known as a gyrator [31], a structure that consists of two transconductors. It is shown in Figure 9.23 and is described by the following transfer function:

$$v_1 = \frac{C}{g_m^2}\frac{di_1}{dt} \tag{9.37}$$

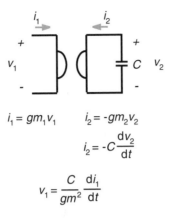

$$i_1 = gm_1 v_1 \qquad i_2 = -gm_2 v_2$$

$$i_2 = -C \frac{dv_2}{dt}$$

$$v_1 = \frac{C}{gm^2} \frac{di_1}{dt}$$

Figure 9.23 Development of gyrator function.

where g_m is the symbol for transconductance,[6] equal to i_{out}/v_{in}.

The utility of the transconductance-C filter is strongly tied to the linearity of the transconductance circuits. The circuit needs to exhibit linear operation over the expected signal range; otherwise, harmonic and intermodulation distortion will be produced. Similarly, circuit noise muddles low-signal operation. To offset thermal noise, the impedance levels of the transconductors and resistors are lowered. But that raises the capacitance, area, and power dissipation of the filter. As a result, dynamic range (S/N), capacitance (C), and power dissipation (P) can be traded off using the following generalized equations [32]:

$$C = \alpha \frac{kTQ}{V_p^2}\left(\frac{S}{N}\right)^2 \tag{9.38}$$

$$P = \eta kTQf\left(\frac{S}{N}\right)^2 \tag{9.39}$$

where α and η are heavily dependent on the filter order, specifications, topology, and active devices; k is Boltzman's constant; T is the absolute temperature; and V_p is the peak signal value.[7] Consequently, the dynamic range of the filter is less than passive LC filters and is a key design consideration.

Transconductance circuits can use either bipolar or FET transistors. Bipolar transistors offer larger transconductance values and wide tuning ranges

---

6. This leads to the shorthand notation, $g_m C$ filters.
7. V_p is assumed proportional to the supply voltage.

because the collector current can be varied with little change in base-emitter voltage. Figure 9.24 illustrates two example circuits, both of which use linearization techniques to expand the useful operating range of the transconductance circuit. Figure 9.25 shows two MOSFET transconductance circuits. In those circuits, MOSFETs are also used to linearize the voltage to current transformation.

Capacitor and transistor variations due to integrated circuit fabrication are minimized by tuning methods. In general, those methods adjust the transconductance of the gyrator circuit to achieve the desired filter response. Furthermore, tuning can be performed once during manufacturing or continuously using a frequency reference. The approaches are detailed in [32–35].

The baseband channel select filters are seventh-order elliptic filters with over 40-dB adjacent channel rejection. The elliptic filter provides the sharpest transition band and the lowest shape factor. As a result, the integrated filter requires the fewest active elements and the lowest power consumption.

Figure 9.26(a) shows the LC prototype of the elliptic filter. In the $g_m C$ filter structure, the inductors are replaced by gyrator-capacitor combinations, as shown in Figure 9.26(b).

The elliptic filter distorts the signal phase near the passband edge and potentially lowers receiver performance. To compensate for that phase non-linearity, the digital modulator in the base station includes a predistortion filter [36].

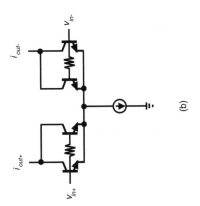

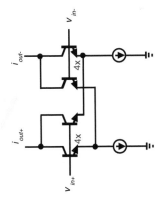

Figure 9.24 Two bipolar-based transconductors.

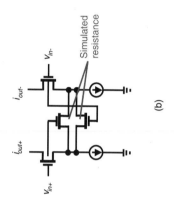

(a)

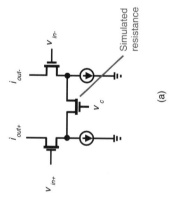

(b)

Figure 9.25 Two MOSFET-based transconductors.

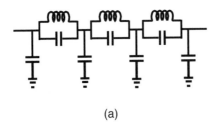

(a)

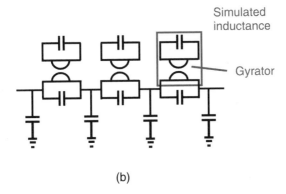

(b)

Figure 9.26 Elliptic filter: (a) LC prototype and (b) transconductance-C implementation.

References

[1]   Gilbert, B., "The Design of Bipolar Si/SiGe LNAs from the Ground Up," BCTM97 Short Course Notes.

[2]   Gonzalez, G., *Microwave Transistor Amplifiers*, 2nd ed. Prentice Hall, 1997.

[3]   Ghapurey, R., and T. Viswanathan, "Design of Front End RF Circuits," *Proc. 1998 Southwest Symp. on Mixed-Signal Design*, pp. 134–139.

[4]   Shaeffer, D. K., and T. H. Lee, "A 1.5V, 1.5 GHz CMOS Low-Noise Amplifier," *IEEE J. Solid-State Circuits*, May 1997, pp. 745–759.

[5]   Abidi, A. A., "High-Frequency Noise Measurements on FET's With Small Dimensions," *IEEE Trans. on Electron Devices*, Vol. Ed-33, Nov. 1986, pp. 1801–1805.

[6]   Wang, B., J. Hellums, and C. Sodini, "MOSFET Thermal Noise Modeling for Analog Integrated Circuits," *IEEE J. Solid-State Circuits*, Vol. 29, July 1994, pp. 833–835.

[7]   Shaeffer, D. K., and T. H. Lee, "A 1.5V, 1.5 GHz CMOS Low-Noise Amplifier," *IEEE J. Solid-State Circuits*, May 1997, pp. 745–759.

[8]   Hughes, B., "A Temperature Noise Model for Extrinsic FETs," *IEEE Trans. on Microwave Theory and Techniques*, Vol. 40, No. 9, Sept. 1992, pp. 1821–1832.

[9]   Hull, C. D., "Analysis and Optimization of Monolithic Downconversion Receivers," University of California—Berkeley, Ph.D. dissertation.

[10] Fong, K. L., C. D. Hull, and R. G. Meyer, "A Class-AB Monolithic Mixer for Downconverter Applications," *IEEE J. Solid-State Circuits*, Vol. 32, Aug. 1997, pp. 1166–1172.

[11] Maas, S., *Nonlinear Microwave Circuits*, IEEE Press, 1988, pp. 267–268.

[12] ibid.

[13] Shahani, A., D. Schaeffer, and T. Lee, "A 12mW Wide-Dynamic Range CMOS Front-End for a Portable GPS Receiver," *IEEE J. Solid-State Circuits*, Vol. 32, No. 12, Dec. 1997, pp. 2061–2070.

[14] Maas, S., *Nonlinear Microwave Circuits*, IEEE Press, 1988, pp. 418–420.

[15] Dura, P., and R. Dikshit, "FET Mixers for Communications Satellite Transponders," *IEEE MTT-S Internat'l Microwave Symp.* Digest, 1976, pp. 90–92.

[16] Pucel, R., D. Masse, and P. Berra, "Performance of GaAs MESFET Mixers at X-Band," *IEEE Trans. on Microwave Theory Tech.*, Vol. MTT-24, June 1976, pp. 351–360.

[17] Dreifuss, J., A. Madjar, and A. Bar-Lev, "A Novel Method for the Analysis of Microwave Two-Port Active Mixers," *IEEE Trans. on Microwave Theory and Techniques*, Vol. MTT-33, 1985, p. 1241.

[18] Tsironis, C., R. Meirer, and R. Stahlman, "Dual-Gate MESFET Mixers," *IEEE Trans. on Microwave Theory and Techniques*," Vol. MTT-32, Mar. 1984, pp. 248–255.

[19] Schmoock, J., "An Input Stage Transconductance Reduction Technique for High-Slew Rate Operational Amplifiers," *IEEE J. Solid-State Circuits*, Vol. SC-10, No. 6, Dec. 1975, pp. 407–411.

[20] Meyer, R. G., "Intermodulation in High-Frequency Bipolar Integrated Circuit Mixers," *IEEE J. Solid-State Circuits*, Vol. 21, Aug. 1986, pp. 534–537.

[21] Razavi, B., "A 1.5V 900 MHz Downconversion Mixer," *ISSCC Digest of Tech. Papers*, Feb. 1996, pp. 48–49.

[22] Netterstrom, A., and E. Christensen, "Correction for Quadrature Error," *Proc. IGARSS94*, pp. 909–911.

[23] McDonald, M., "A 2.5 GHz BiCMOS Image-Reject Front-End," *ISSCC Digest Tech. Papers*, Feb. 1993, pp. 144–145.

[24] Zverev, A. I., *Handbook on Electrical Filters*, New York: Wiley, 1967.

[25] Williams, A. B., and F. J. Taylor, *Electronic Filter Design*, New York: McGraw-Hill, 1995.

[26] Lindquist, C. S., *Active Network Design With Signal Filtering Applications*, Long Beach, CA: Steward, 1977.

[27] Razavi, B., "CMOS RF Receiver Design for Wireless LAN Applications," *IEEE Proceedings of Radio and Wireless Conference*, Aug. 1999, pp. 275–280.

[28] Tsividis, Y. P., and J. O. Voorman, eds., *Integrated Continuous-Time Filters*, New York: IEEE Press, 1993.

[29] Tsividis, Y., and P. Antognetti, eds., *Design of MOS VLSI Circuits for Telecommunications*, Englewood Cliffs, NJ: Prentice Hall, 1985.

[30] Gregorian, R., and G. C. Temes, *Analog MOS Integrated Circuits for Signal Processing*, New York: Wiley, 1986.

[31] Tellegren, B. D. H., "The Gyrator, a New Electric Network Element," *Philips Research Reports*, Vol. 3, 1948, pp. 81–101.

[32] Tsividis, Y. P., "Integrated Continuous-Time Filter Design—An Overview," *IEEE J. Solid-State Circuits*, Vol. 29, No. 3, Mar. 1994, pp. 168–176.

[33] Schaumann, R., and M. A. Tan, "The Problem of On-Chip Tuning in Continuous-Time Integrated Filters," *IEEE Proc. ISCAS*, 1989, pp. 106–109.

[34] VanPeteghem, P. M., and R. Song, "Tuning Strategies in High-Frequency Integrated Continuous-Time Filters," *IEEE Trans. on Circuits and Systems*, No. 1, Jan. 1989, pp. 136–139.

[35] Kwan, T., and K. Martin, "An Adaptive Analog Continuous-Time CMOS Biquadratic Filter," *IEEE J. Solid-State Circuits*, Vol. SC-26, No. 6, June 1991, pp. 859–867.

[36] TIA/EIA Interim Standard, "Mobile Station-Base Station Compatibility Standard for Dual-Mode Wideband Spread Spectrum Cellular System," IS95a, Apr. 1996.

10

Next-Generation CDMA

As the demand for wireless services grows, new methods of delivery with greater access and higher data rates are needed. That in turn requires more efficient use of the limited radio resources.

The goal of third-generation (3G) communications is to provide high-speed data with reasonable capacity, while improving multipath resolution and increasing diversity [1, 2]. The target data rates are 144 Kbps for wide-area usage and full mobility, 384 Kbps for urban use, and up to 2 Mbps for virtual home service and low mobility. By comparison, the data rates for 2G networks typically are under 20 Kbps. The higher data rates will increase voice capacity, enable video communications, and introduce a slew of advanced digital services. They will enable user-friendly access to the Internet, short messaging services (SMS) with embedded photographs or video clips, video telephony, location-based services, and other yet-to-be-imagined services.

To deliver those services efficiently, packet-data and packet-switched connections are required, as opposed to existing circuit-switched connections [1, 3]. Such a delivery method utilizes variable spreading rates, "bundled" multiple data channels, and new spreading techniques. It also means changes to the physical and logical channels, the network architecture, and the protocol stack of 2G systems.

The 2G systems look to evolve to a single worldwide standard known as the Third Generation Partnership Project (3GPP) [4]. That single standard provides the potential for global communications by combining three options in a single framework. It consists of two single carrier options,[1] which receive support from the Association of Radio Industry and Business (ARIB) in Japan

---

1. These options were originally proposed as WCDMA and TDMA/CDMA.

and the European Telecommunications Standards Institute (ETSI). It also lists a multicarrier option,[2] which evolves directly from CDMA IS95 and draws support from the Telecommunications Industry Association (TIA) subcommittee TR45.5. Interestingly, all three options embrace direct-sequence spread-spectrum CDMA technology.

This chapter presents an overview of next-generation CDMA communication systems described by the 3GPP standard. It introduces several key concepts, which have evolved from the CDMA IS95 system, to provide greater access and higher data rates. It also reviews the three options for next-generation wideband CDMA systems, single carrier with FDD operation, single carrier with TDD arrangement, and multicarrier modulation.

10.1 Concepts of Next-Generation CDMA

Next-generation services will require improved network efficiency to provide greater access and to deliver high-speed data. Second-generation networks limit access through hard factors, such as time and frequency, that divide the radio spectrum and circuit-switched connections that limit data throughput. These networks assign fixed radio channels and dedicated network paths for each user. In contrast, 3G networks introduce soft factors and packet-switched connections that are dynamic and thus more efficient.

A packet-switched network assigns resources based on the data throughput demands and the quality of service (QoS) requirements of each application. For example, voice services rely on low data rates and limit processing delay. In contrast, Internet browsers and multimedia applications work with large data bursts and can tolerate greater delay. Therefore, the two application types require different resources from the network.

The network efficiently maps the available resources to the needs of each application using improvements in several key areas, including frequency diversity, flexible data rates, spreading techniques, source or error correction coding, and reverse-link coherent detection, which are outlined next.

10.1.1 Next-Generation CDMA and the Physical Channel

A wide bandwidth radio channel is used to deliver the high data rates needed for advanced 3G services. It is nominally 5 MHz wide (that is, 99% of the radio energy is contained in a 5-MHz bandwidth) and is formed by one of two methods, depending on the available radio spectrum. In many places in

---

2. This option was first proposed as cdma2000 [5].

the world, new radio spectrum, known as the IMT-2000 radio bands, is available, as shown in Figure 10.1. However, in some countries like the United States, this radio spectrum is not available, so reuse of the cellular and PCS radio spectrum is needed. In other countries such as China, only a portion of the new radio spectrum is available; thus, TDD operation is planned. Such a scattered frequency plan presents a major obstacle for global communications.

The wide bandwidth signal is generated by direct sequence spreading of one or multiple carriers. In the single-carrier approach, the spreading chip rate is increased to form a wideband signal, as shown in Figure 10.2(a). If the forward and reverse links share the same frequency channel, TDD operation is also employed. It divides the channel into time slots and alternately assigns the slots to the forward and reverse links. In the multicarrier approach, the wideband signal is formed by three narrowband, contiguous direct-sequence signals, as shown in Figure 10.2(b). The advantage of that approach is that each narrowband signal is compatible with CDMA IS95-modulated signals. That concept is essential to 3GPP systems that overlay existing 2G networks.

The wider bandwidth signal provides improved frequency and multipath diversity. Flat frequency fading occurs less often with a true wideband signal. Instead, a portion of the signal experiences frequency-selective fading. The high-speed spreading sequence also results in a shorter chip period. That provides sharper resolution of the cross-correlation function in the searcher, which leads to better channel estimation and phase synchronization in the Rake receiver [3] but requires a more complicated receiver.

10.1.2 Multirate Design in Next-Generation CDMA

One of the motivations for 3G communication systems is increased flexibility, which is introduced by packet data and multirate operation. Packet data allows the connection to adapt to varying application requirements and, in some way, enables the network to serve as many users as possible. Multirate operation supports different data rates through variable spreading factors and multiple spreading codes.

Multirate design maps variable-width data packets, also known as transport blocks, to fixed-length data frames. The number of bits in each data packet is linked to the application but is limited to a set of defined values. In contrast, the data frame is constant, typically 10 or 20 ms in length, and holds a fixed number of chips.[3]

The data packet is mapped to the fixed frame using two basic techniques: variable spreading factors and multiple codes. In the first method, the data

---

3. The fixed data frame contains a defined number of chips, which produces the designated modulation bandwidth.

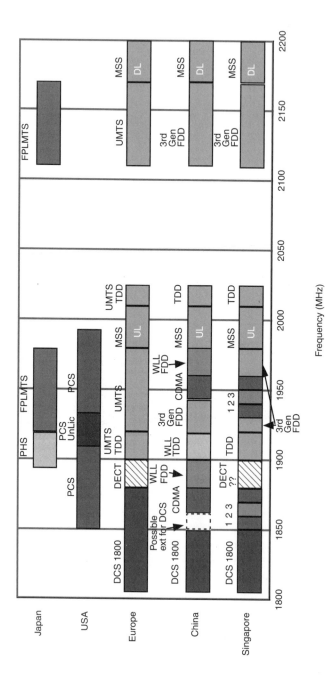

Figure 10.1 Worldwide frequency plan for next-generation services.

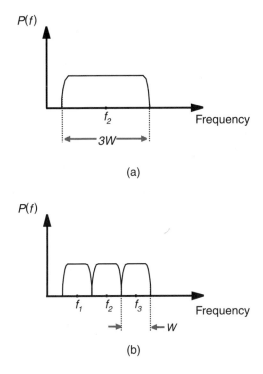

Figure 10.2 Proposed modulation schemes for next-generation CDMA: (a) single-carrier and (b) multicarrier.

packet is spread to the size of the fixed frame by a variable spreading code. The spreading factor, W/R, is given by

$$\frac{W}{R} = \frac{BT_F}{N} \tag{10.1}$$

where B is the spread-spectrum modulation bandwidth (3.84 Mcps for the single-carrier options), T_F is the frame length, and N is the number of bits in each data packet. The spreading factor W/R ranges from 4 to 256 and describes the processing gain of the system. In the second method, the data frame is formed by M parallel channels with fixed capacity. Therefore,

$$M = \frac{N}{RT_F} \tag{10.2}$$

where R is the bit rate of the single channel. Furthermore, it is also possible to combine the two methods.

The size and the frequency of the transport blocks fluctuate with most applications. Ideally, the digital modulator would follow the changes, adjusting the spreading factor and/or the number of coded channels instantly, thereby providing the needed data throughput. In practice, that is impractical because the changes affect the network and, more important, affect other users. Managing packet data and packet-switched connections requires network scheduling via the radio resource control (or network) protocol layer. That protocol layer defines transport block sets that cover 10-, 20-, 40-, or 80-ms intervals.

Figure 10.3(a) is an example of a transport block set for voice service. It contains one data packet, which repeats every 10 ms, and provides a continuous connection. Figure 10.3(b) shows a second example, in which two transport blocks share the channel. Voice data packets fill every other slot, while data bursts occur during the second and fourth slots. This transport block set repeats every 80 ms.

These multirate approaches affect the received bit energy per noise density ratio (E_b/N_o), a key communication-link parameter. That is because the spreading factor alters the bit interval and thus changes the received bit energy per noise density ratio in the following way

$$\frac{E_b}{N_o} = \left(\frac{S}{I}\right)\frac{W}{R} \qquad (10.3)$$

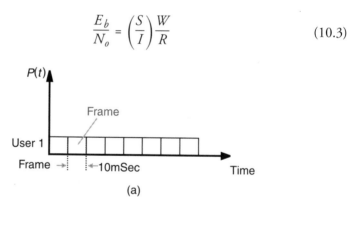

(a)

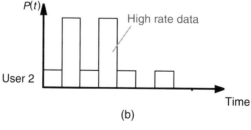

(b)

Figure 10.3 Examples of transport block sets: (a) voice only and (b) voice plus data.

where S is the desired signal power and I is the interference power, which includes the power from other users plus thermal noise. To obtain the same value of E_b/N_o and thus similar link performance, the power allocated to the wanted signal or the performance of the receiver must change. Adjusting the receiver is impractical, so the link performance is optimized by adjusting the transmit signal power to keep E_b/N_o roughly constant.

Although the resulting transmit power levels are unequal, the near-far problem is not introduced, which the following example illustrates. Two users are running different applications; the first user is connected to a voice call at 15 Kbps, while the second user is connected to an Internet application at 120 Kbps. Each user fills a fixed 10-ms data frame and spreads the data to 3.84 Mcps. To do that, the first user applies a spreading factor of 256, and the second user applies a factor of 32. By design, the transmit power levels of each user are selected to achieve equal E_b/N_o values. That means

$$256\left(\frac{S}{I}\right)_1 = 32\left(\frac{S}{I}\right)_2 \tag{10.4}$$

The transmitted signal power levels allocated to the first and second user are set to S and αS, respectively. In this example, the other user is assumed to be the dominant source of interference, therefore $I_1 = \alpha S$ and $I_2 = S$. Substituting into (10.4) yields $\alpha = 2.74$ and means the power of the second user, S_2, is optimally set 4.3 dB higher than the power of the first user, S_1.

The benefits of packet-switched connections are many. The connections introduce a soft factor, namely, power, to divide the radio resource. Because of that, it is now possible to efficiently share the communication link among multiple users with different data requirements, as shown in Figure 10.4. The radio resource control protocol is responsible for multirate operation and transmit power management [5].

10.1.3 Spreading Technique for Next-Generation CDMA

In CDMA communications, the spreading codes are crucial. They allow synchronization to the network and provide the means for multiple access. Spreading codes have additional requirements with multirate design, including better correlation performance to improve synchronization and receiver performance, a higher number of orthogonal codes to accommodate more users, and greater code flexibility to handle variable spreading factors.

Pseudorandom noise (PN) sequences are used for synchronization because these signals appear noiselike and demonstrate excellent autocorrelation features. A typical PN sequence is an M-sequence, which is generated by an M-bit

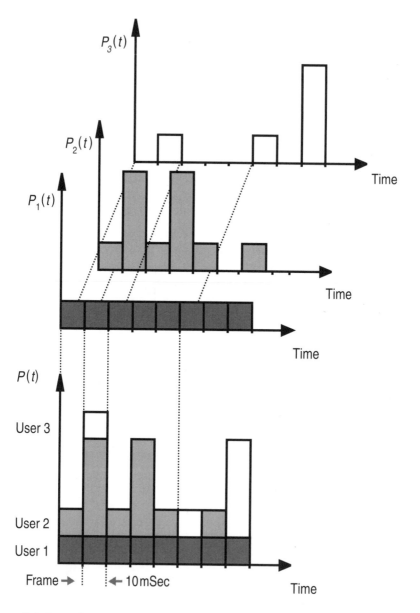

Figure 10.4 Power levels in a multirate system.

linear-feedback shift register. Each offset of the M-sequence is in fact a new sequence with good autocorrelation properties. In practice, the usefulness of M-sequences is limited by excess delay variations caused by the channel and partial-correlation results produced in the demodulator.

Spreading codes also need to separate users. That requires deterministic codes with good cross-correlation properties. M-sequences become less orthogonal with signal offsets, variable spreading factors, and asynchronous networks, limiting their usefulness [6].

The Walsh-Hadamard recursive technique is an example of an algorithm that generates orthogonal treelike codes, as shown in Figure 10.5 [1, 6, 7]. These codes are mutually orthogonal and are useful for variable-length spreading when certain rules are followed. The rules prevent the use of codes from the same code-tree path. That is important because the shorter-length codes are used to construct the longer-length codes and thus have potentially poor cross-correlation properties [7].

There are other types of spreading codes, including Gold [8] and Kasami codes [9]. These codes demonstrate good aperiodic autocorrelation properties and deliver better cross-correlation performance than M-sequences. As such, these codes are especially useful in asynchronous networks.

The pseudorandom and orthogonal codes are used to modulate the message signal to the wide spread-spectrum bandwidth. CDMA IS95 uses the balanced quaternary spreading technique shown in Figure 10.6(a). It spreads the message data using two orthogonal high-rate sequences. This approach actually duplicates the message data to the two orthogonal signals.

To double the data rate, the message signal can be split into two independent streams prior to spreading. This is shown in the dual-channel QPSK spreading circuit in Figure 10.6(b). To reduce the amplitude variation of the modulated signal's envelope, a complex spreading technique is introduced. It uses a complex sequence or two real sequences and is illustrated in Figure 10.6(c).

Table 10.1 compares the characteristics of the spreading techniques. The advantage of complex spreading—the choice for 3G communication systems—

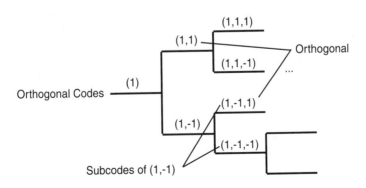

Figure 10.5 Code tree structure.

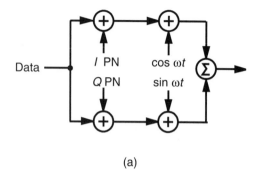

(a)

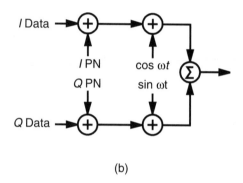

(b)

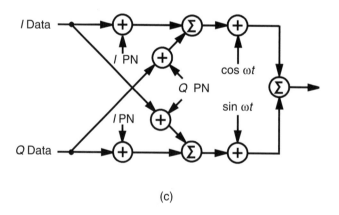

(c)

Figure 10.6 QPSK spreading techniques: (a) balanced quarternary spreading, (b) dual-channel QPSK spreading, and (c) complex QPSK spreading.

Table 10.1
Characteristics of Spreading Circuits Applied to a Single-Coded Channel

Spreading Technique	Modulation	Data Rate	Envelope Variation
Balanced quarternary	QPSK		5.6 dB
	OQPSK		5.1 dB
Dual channel	QPSK	2x	5.7 dB
Complex spreading	QPSK	2x	4.1 dB

is clear. That advantage grows with multicoded channels, which can have envelope variations as large as 8 to 10 dB even with complex spreading.

10.1.4 Advanced Error Control Techniques for Next-Generation CDMA

A key benefit of digital communications is the robustness provided by error control methods. Error control methods combat radio propagation effects and allow the system to operate at lower transmit power levels. In general, error control methods fall into one of two categories: forward error correction (FEC), which provides data protection, and handshake protocols, which facilitate detection of corrupted data and retransmission requests. The choice of error control method is linked to the application: Low latency and moderate BER (below 10^{-3}) are needed for voice communication, while longer delay is tolerated but lower BER (near 10^{-6}) is desired for data transmission.

Convolutional codes are a common FEC method, with moderate coding gains, and are suitable for voice communications because of their low latency and low complexity [10]. Convolutional codes are described by their code rate and constraint length.

A new class of convolutional codes, turbo codes [11, 12], provides an alternative FEC technique. Turbo codes improve the reliability of communication links and amazingly approach the channel capacity (in AWGN) predicted by Shannon [11, 13]. As such, turbo codes are ideal for data communications with low BER requirements.

In general, two kinds of convolutional encoders are of practical interest: nonsystematic convolutional (NSC) coders and recursive systematic convolutional (RSC) coders. The nonsystematic convolutional encoder was presented in Section 5.1.2. The recursive systematic convolutional coder is actually obtained from the nonsystematic convolutional encoder by using feedback and setting one of two outputs equal to the input data. The two encoders, shown in Figure 10.7, are capable of similar error correction performance (i.e., equivalent

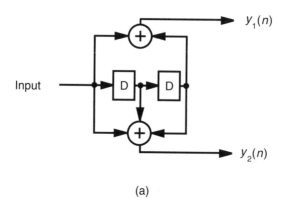

(a)

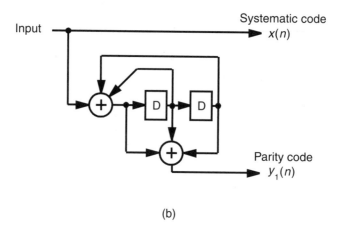

(b)

Figure 10.7 Convolutional encoders: (a) NSC and (b) RSC.

minimum free distance), although the RSC provides better performance at low SNR levels [12].

A turbo encoder joins a systematic code with two parity codes, generated by RSC encoders, as shown in Figure 10.8(a). The two RSC encoders are connected in parallel and separated by a nonuniform interleaver to dramatically lower the probability of error at high data rates. The interleaver function, not the constituent RSC encoders, actually sets the performance of the turbo encoder. Ideally, the nonuniform interleaving provides maximum scattering of the data, increasing the minimum free distance of the code and making the two redundant data streams as diverse as possible [12, 14]. Finding the optimum interleaving function is the real challenge in turbo code design.

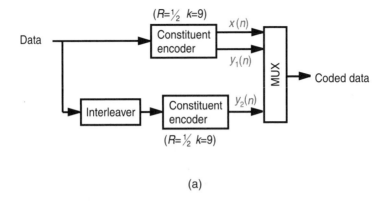

(a)

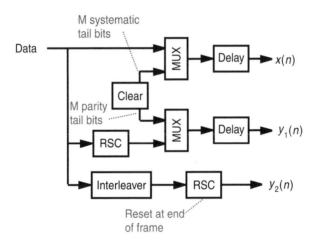

(b)

Figure 10.8 Turbo encoder provides improved coding gain compared to conventional codes: (a) simplified view and (b) complete view.

The turbo code is constructed as follows. The input data flows directly to the first constituent encoder and, in parallel, feeds an interleaver and a second constituent encoder. Delay is added to the systematic (uncoded) data and the output of the first encoder to align all three output data streams. As with the NSC encoder, the turbo encoder is "flushed" after each frame to provide a predetermined starting point for the decoding algorithm and to prevent carryover or memory from one data frame to the next.

The benefits of turbo codes are due primarily to the iterative decoding algorithm. It uses a maximum a posteriori scheme based on the BCJR algorithm [15]. Note that the Viterbi algorithm, which is a maximum likelihood decoding scheme used for NSC codes, is not optimal for turbo codes.

The turbo decoder, shown in Figure 10.9, uses suboptimal, soft-decoding rules that decode each RSC code separately. Furthermore, it shares those results in an iterative fashion to extract the original data [16].

Each RSC code is decoded starting from the end of the frame and moving backward, similar to the Viterbi algorithm. The decoding procedure produces both intrinsic data and extrinsic data. The extrinsic data, which is unavailable in the Viterbi decoder, is crucial to the turbo decoding process because it prevents information produced by the first decoder and passed to the second decoder from being fed back to the first decoder. As a result, it diversifies the interative decoding process [11].

Recall that the Viterbi algorithm produces a log-likelihood function (see Section 5.2.5) for the decoding path defined by

$$\Lambda(\mathbf{d}|\mathbf{x}) = \ln p(\mathbf{d}|\mathbf{x}) = \sum_{all \, n} \ln p(d(n)|x(n)) \qquad (10.5)$$

where $\mathbf{x}$ is the input vector and $\mathbf{d}$ is the decoded output vector. In a binary system, where the $d(n)$ is either 0 or 1, a related function—the log-likelihood ratio—is useful. Here,

$$\Lambda(d(n)) = \log \frac{p[d(n) = 1|x]}{p[d(n) = 0|x]} \qquad (10.6)$$

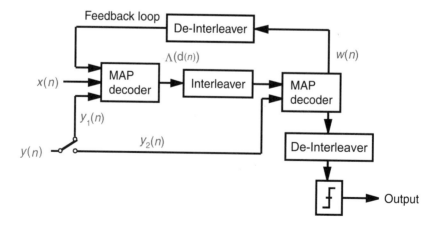

Figure 10.9 Block diagram of turbo decoder.

where $p[d(n)|x]$ is the a posteriori probability. In contrast to the Viterbi decoder, the maximum a posteriori probability decoder produces an output for each input bit [11, 12, 14]. As a result, the log-likelihood ratio can be factored into the following expression [11]:

$$\Lambda[d(n)] = \Lambda[d(n)|x]\frac{2}{\sigma^2}x(n) + w(n) \qquad (10.7)$$

where σ^2 is the variance of the received noise and $w(n)$ is referred to as the "extrinsic" information. The extrinsic information is important because it is independent of the input data $x(n)$.

The turbo decoder iteratively processes the data until an error criterion is met. In practice, the first few iterations provide the greatest performance gain, as shown in Figure 10.10. Note that additional iterations further burden the DSP and increase latency in the data path.

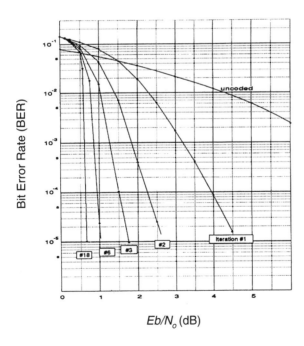

Figure 10.10 The benefit of iterative decoding as applied to turbo codes (*From:* C. Berrou and A. Glavieux, "Near Optimum Error Correcting Coding and Decoding: Turbo Codes," *IEEE Trans. on Communications,* © 1998 IEEE).

The second category of error control is based on error detection methods instead of data protection techniques [5]. It uses a parity or error detecting code to verify each data frame. If an error is detected, a repeat request is sent and the frame is retransmitted. Because error detection is far simpler than data protection, this technique is extremely efficient.

A common protocol of this type is the automatic repeat request (ARQ) scheme [17]. It and other handshake protocols are useful only for data services because any retransmission delays are unacceptable for voice communications.

10.1.5 Coherent Detection Methods

Coherent data detection of phase-modulated signals requires a reference signal that can be sent by the data source or reconstructed from the received data. In practice, it is often difficult to reconstruct the reference signal, especially in wireless communications that are subjected to fading effects. When a reference signal is transmitted, the performance benefit can be as much as 3 dB. However, the performance gain is significantly lower when the reference estimate is poor. Furthermore, differential modulation schemes, such as $\pi/4$DQPSK, can provide performance similar to coherent detection methods.

The phase reference signal is known as the pilot signal and is transmitted either continuously or is multiplexed into the data stream, as illustrated in Figure 10.11. These approaches provide different benefits: The continuous pilot is immune to fast fading, while the multiplexed pilot is better at minimizing self-interference. In either case, the effectiveness of the pilot is based on the transmitted power level.

10.1.6 Interoperability in Next-Generation CDMA

The three options for next-generation CDMA are not compatible at either layer 1 (the physical layer) or layer 2 (the MAC and radio link control layers) due to different modulation methods, physical channel designs, and logical channel formats.

The single-carrier options (FDD and TDD) provide an upgrade path for GSM and primarily interface with ISDN core networks using the mobile application part (MAP) protocol [5]. The multicarrier option mirrors CDMA IS95 and connects to the telephone network using the IS41 network protocol [18].

These network protocols contain common attributes, namely notification and control of the radio resources, QoS messages, and reporting information. As such, it is possible for the three options to be compatible at layer 3 (the network level). In fact, the 3GPP standard provides for interoperability at that

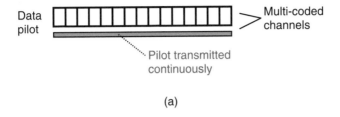

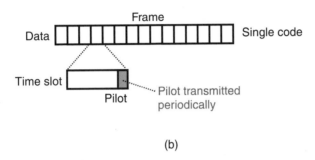

Figure 10.11 Pilot signal concepts: (a) continuous and (b) multiplexed.

level and maps the different radio link control layers to the different network protocols.

10.2 Single-Carrier CDMA Option

The evolution path for TDMA systems, such as GSM and NADC, leads to the single-carrier options of the 3GPP standard. For GSM systems, that path includes high-speed circuit-switched data (HSCSD) networks [19, 20] at data rates of 57 Kbps and general packet radio protocol system (GPRS) networks [21, 22] with maximum data rates of 170 Kbps. These intermediate steps enhance the capabilities of 2G systems and lessen the urgency for 3G systems.

The 3GPP single-carrier option borrows the frame structure and protocol stack from the GPRS enhancement. That ensures backward compatibility of the single-carrier options at the higher protocol layers but cannot mitigate differences at the physical layer and thus requires new radio spectrum.

The single-carrier options use a nominal chip rate of 3.84 Mcps for direct sequence spreading [2].[4] The FDD single-carrier option also defines logical

---

4. The original chip rate was 4.096 Mcps, but it has been standardized in the 3GPP proposal to 3.84 Mcps.

channels, generates data frames, and approaches synchronization differently than CDMA IS95 systems, as outlined below.

10.2.1 Forward Link in the Single-Carrier Option

The single-carrier forward-link modulator is depicted in Figure 10.12. Data is FEC encoded and mapped to 10-ms frames using static rate matching. Rate matching repeats or punctures symbols to achieve the designated number of bits per frame requested by the service and scheduled by the network. The data frame is then interleaved to provide time diversity. Multiple services or high-speed data are then multiplexed to a single dedicated channel, as specified by the structure of the transport block set. The multiplexed data stream is adjusted to fit rates supported by the network using dynamic rate matching. Each transport block is then interleaved, channelized, and scrambled using orthogonal variable spreading factors (OVSFs), or *long* codes. Root-raised cosine filters, with a roll-off factor (α) equal to 0.22, are used to limit the transmitted spectrum to a nominal 5-MHz bandwidth [3]. Finally, the data is QPSK-modulated using *short* PN sequences.

The forward link of the single-carrier FDD option consists of several logical channels, as shown in Table 10.2. These logical channels provide familiar functions with increased capacity and flexibility.

The broadcast channel originates from the base station and communicates information to the cell area or to the entire network. This logical channel shares the radio resource with the sync channels in a TDMA scheme, as shown in Figure 10.13. The sync channels are transmitted at the start of each slot and are designed to coordinate timing in the network.

The forward-access channel carries control information and short user packets within the cell boundary. It can be used to transport short bursts of data without establishing a new dedicated data channel or modifying an existing channel.

The paging channel carries control information to a mobile with an unknown location. A short, uncoded message, known as a paging indicator signal, indicates whether the paging channel needs to be decoded.

The control and data logical channels combine to form a dedicated physical channel, as shown in Figure 10.14. Each physical channel maps to 10-ms frames with 15 time slots. Each time slot includes the transport format combination indicator (TFCI), transport power control (TPC), optional pilot symbols, and data. The exact number of bits assigned to each field is based on several factors, including overall data rate. In general, the number of bits allocated for control is a small fraction of the overall data rate and is referred to as overhead.

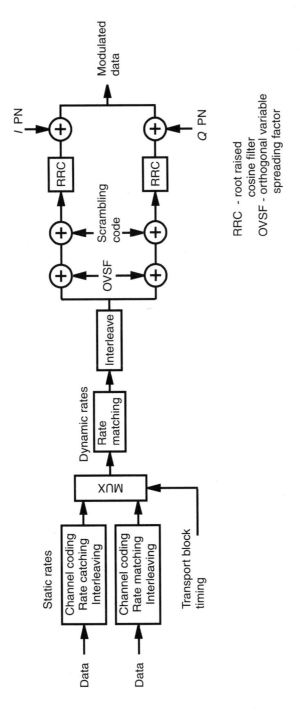

Figure 10.12 Block diagram of the forward-link modulator used in single-carrier CDMA.

Table 10.2
Forward-Link Parameters of Single-Carrier FDD Options [23, 24]

Channel	Data Rate (Kbps)	Channel Coding	Processing Gain
Pilot	—	None	—
Broadcast	30	Rate 1/2	256
Primary sync	256 chips, 1/10 slot, once per frame		—
Secondary sync	256 chips, 1/10 slot, repeat every slot		—
Access	16	Rate 1/2	240
Paging	16	Rate 1/2	240
Dedicated control	15–1,920	Rate 1/22	N = 512 to 4
Dedicated data		Turbo	

The TFCI specifies the number of bits, N, in each time slot using this relationship:

$$N = 10 \times 2^k \qquad (10.8)$$

where N is limited to the set defined by $k = 0, 1, 2, \ldots, 6$. The value of k corresponds to a spreading rate (W/R) given by

$$\frac{W}{R} = \frac{256}{2^k} \qquad (10.9)$$

If the spreading rate is known, so are the symbol repetition and puncture rates. As a result, the rate information is readily available and rate determination is avoided. Note that the TFCI is not used for fixed rate services.

The TPC instructs the mobile telephone to decrease or increase its output power level. This feedback signal is used in the closed-loop power control algorithm.

Pilot symbols are inserted in each time slot to provide a dedicated pilot signal, which is used to augment the common pilot signal. This technique is especially effective in adaptive antenna arrays and allows more efficient closed-loop power control.

10.2.2   Reverse Link of Single-Carrier Option

Figure 10.15 illustrates the reverse-link modulator, which performs the same basic operations as the forward-link modulator. It uses dual-channel QPSK

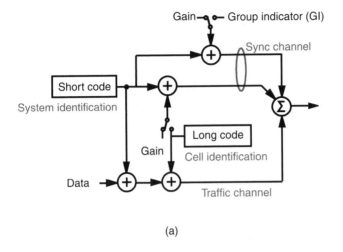

(a)

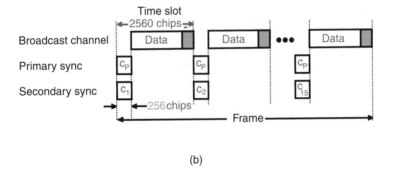

(b)

Figure 10.13 Time slots for broadcast primary and sync channels: (a) generation and (b) timing.

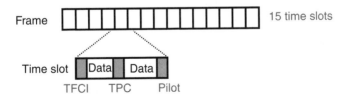

TFCI - transport channel format indicator
TPC - transport power control

Figure 10.14 Dedicated channel frame structure for proposed 3GPP single-carrier FDD option.

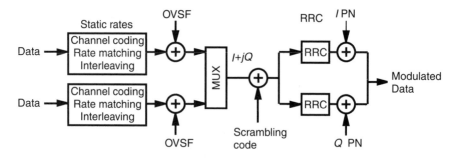

Figure 10.15 Block diagram of the reverse-link modulator used in single-carrier CDMA.

modulation and bundles additional channels in the multiplexer, differentiated by orthogonal codes, to achieve higher data rates. The reverse-link modulator uses the same OVSF codes as the forward-link modulator to ensure orthogonality and thereby minimize interference.

The reverse link consists of several logical channels, which are listed in Table 10.3 and outlined next.

The access channel initiates communications and responds to messages sent on the broadcast, forward access, or paging channels. It is shared by all the mobiles in the cell coverage area and is characterized by a risk of collision, which occurs when more than one mobile radio requests service at the same time.

The common packet channel transports small- and medium-sized data packets that complement the data capabilities of the forward-access channel.

Table 10.3
Reverse-Link Parameters for Proposed 3GPP Single-Carrier FDD Option [23, 24]

Channel	Data Rate (Kbps)	Channel Coding	Processing Gain
Pilot	—	None	—
Access			
Data	15–120	Rate 1/2	N = 256 to 32
Control	15	Rate 1/2	256
Common packet	30	Rate 1/2	256
Dedicated control	15–960	Rate 1/2	N = 256 to 4
Dedicated data		Turbo	

It is a common channel, and as such, access is random and contention based. It is strictly intended for burst traffic.

The control and data logical channels combine to form a dedicated physical channel, as shown in Figure 10.16. Instead of time multiplexed (as in the forward-link modulator), the channels are code multiplexed and applied to different arms of a dual-channel QPSK modulator. The values for the number of bits per time slot, N, and the spreading rate, W/R, are found using (10.8) and (10.9).

The exact number of bits for the control channel—consisting of the pilot symbols, TFCI, feedback indicator (FBI), and TPC—is not yet defined. In general, the control channel and the data channel will have different data rates and spreading factors. Furthermore, the number of bits allocated to the control fields will be very small compared to the data fields.

10.2.3 Acquisition and Synchronization

Synchronization occurs at three levels: slot, frame, and scrambling code. The 3GPP single-carrier options promote a network asynchronous[5] scheme to allow continuous operation between indoor and outdoor environments. In those networks, the base stations are not synchronized to each other and are independent of external timing, such as the GPS system [25]. As a result, the problem of synchronization in indoor networks, where GPS timing is unavailable because of weak signals, is avoided. That freedom comes with a price: It affects code synchronization, cell acquisition, and handover. It also means the spreading codes must be effective (i.e., have low cross-correlation properties) even when offset or delayed.

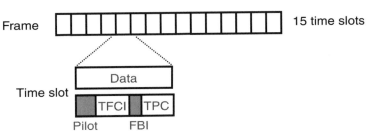

Figure 10.16 Reverse-link frame structure for proposed 3GPP single-carrier FDD option.

---

5. This is different from systems based on CDMA IS95 and the 3GPP multicarrier option, which utilize base stations that are synchronized to each other.

The synchronization process in asynchronous networks involves two code sequences: the primary sync sequence and the secondary sync sequence. The primary sync code (PSC) indicates slot timing, while the secondary sync code (SSC) provides frame timing and the scrambling code.

The PSCs and SSCs are cyclic codes [26, 27]. Cyclic codes provide the feature whereby a cycle or phase shift of the code forms another code word. In other words, the code demonstrates good aperiodic autocorrelation properties. An M-sequence is an example of a cyclic code [28, 29]. Another example is the Golay code [30], which is used to form the PSCs and SSCs, as shown in Figure 10.17(a).

The PSC is formed by modulating a known 16-chip sequence, $a(n)$, by an 8-chip Golay complimentary sequence, $b(n)$. The result is repeated and masked by a 16-bit sequence $y(n)$. Each bit of the sequence $y(n)$ operates on a 16-chip segment; the masking operation either passes or inverts the segment data, as shown in Figure 10.17(b). The result is the PSC $c_p(n)$, which has a length equal to 256 chips. The uniqueness of the code is due to the Golay complimentary sequence, $b(n)$, and is different for each system. The primary sync channel repeats each time slot and, once decoded, provides time slot timing.

The SSC is formed by taking the sequence $b(n)$ and adding it, using modulo-2 arithmetic, to a Hadamard sequence $h_k(n)$. The Hadamard sequences are rows of a 256-by-256 Hadamard matrix, indicated by the index k, and span 256 chips. As a result, the SSC $c_{sk}(n)$ also has a length equal to 256 chips and a period of one time slot. The index of the Hadamard sequence, k, is different for each time slot within a frame, limited to every eighth row of the Hadamard matrix, and restricted to the first 17 indexes (i.e., rows 0, 8, 16, ..., 136). These indexes follow one of 32 patterns and indicate frame timing and the scrambling code.

The acquisition time for an asynchronous network generally is longer than for a synchronous network. That is because in the asynchronous network, both the code phase and the associated chip timing are unknown at the receiver; hence, a two-dimensional search space (time and code) is needed. By contrast, in synchronous networks like CDMA IS95, the M-sequence is known and only the time space is searched.

10.2.4 Fast Power Control

The 3GPP single carrier option implements open-loop and closed-loop power control schemes. As with the CDMA IS95 system, the open-loop technique estimates the forward-link path loss, while the closed-loop technique adjusts the received signal strength at the base station to equalize the signal-to-interfer-

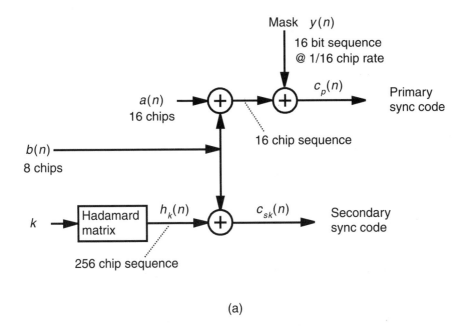

(a)

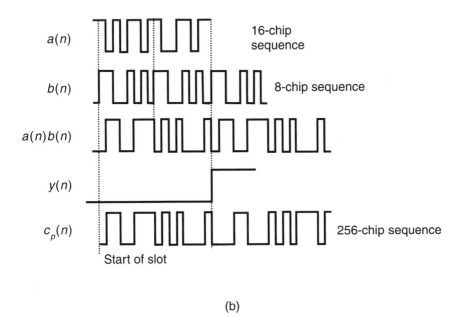

(b)

Figure 10.17 PSCs and SSCs: (a) block diagram of code construction algorithm and (b) details of primary-code generation.

ence ratio of all the mobile telephones in the cell coverage area. As such, the closed-loop technique ensures the quality of the communication links while it maximizes system capacity.

The TPC information is used for closed-loop power control. It is transmitted every time slot (0.667 ms), the equivalent of 1,500 Hz, which is nearly twice as often as in CDMA IS95 networks. In addition, multirate design demands more abrupt changes in transmit power level, for example, when the spreading rate changes from 256 to 4.

10.2.5   Air Interface for the Single-Carrier Option

The single-carrier option is incompatible with 2G TDMA and CDMA systems. As such, it requires new, dedicated radio spectrum, designated the IMT-2000 bands (1,920–1,980 MHz and 2,110–2,170 MHz). The paired frequency bands allow for frequency duplex operation.

The system provides for up to four forms of diversity to improve the wireless link. Channel coding and interleaving provide time diversity to combat burst errors. Wide spread-spectrum signals enable multipath diversity, thereby reducing the effects of small-scale fading. Multiple receive antennas at the base station (and possibly at the mobile radio) provide spatial diversity and also address small-scale fading effects. Supplemental transmit signals from the base station can introduce transmit diversity if deployed and further reduce the impact of small-scale fading. These network features lead to the 3GPP system performance parametrics described in Table 10.4 and Table 10.5.

The single-carrier options measure modulation accuracy with the error vector magnitude (EVM) technique [31], which is defined as

Table 10.4
Minimum Performance Parameters for Mobile Radio Receiver Designed for Single-Carrier FDD Option

Parameter	Condition	Requirement
Sensitivity	BER < 10^{-3}	−110 dBm
Maximum input	BER < 10^{-3}	−25 dBm
Blocking	Adjacent channel @ −30 dBm BER < 10^{-3}	−107 dBm
IMD	Adjacent channel @ −46 dBm BER < 10^{-3}	−107 dBm

Table 10.5
Minimum Performance Requirements of Mobile Radios Designed for
Single-Carrier FDD Option

Parameter	Condition	Requirement
Maximum RF level (class II)		+23 dBm
Minimum controlled RF level		−44 dBm
Adjacent channel power	5 MHz offset	−32 dBc
	BW = 4.096 MHz	−50 dBm
Transmit modulation accuracy		EVM < 17.5%

$$EVM = \frac{\sqrt{\sum_n |e(k)|^2}}{n} \qquad (10.10)$$

where $e(k)$ is the vector error between the actual signal and the ideal symbol and n is the range of symbols, equal to a time slot or frame.

10.3 TDD CDMA Option

The TDD option[6] makes possible 3GPP networks in regions without paired frequency bands or with limited radio spectrum. It uses the same 5-MHz radio channel for both the forward link and the reverse link.

In a TDD system, dividing the radio spectrum into time slots forms the forward- and reverse-link channels. These time slots contain spread-spectrum modulated data that matches data found in the FDD option. As such, each time slot is capable of high data rates. The time slots can be assigned in a flexible way that supports asymmetric links, as shown in Figure 10.18. This scheme is also known as TDMA/CDMA.

Only the dedicated channels are time multiplexed. All other channels, both transport and physical, are transmitted continuously and are identical to those found in the single-carrier FDD option. Furthermore, the mobile radio must receive and demodulate at least two time slots per frame to maintain closed-loop power control.

The TDD option also supports opportunity-driven multiple access (ODMA) operation. In that mode, the time slots are also used as relay slots between base stations. If ODMA is utilized, at least two slots are needed, one

---

6. PHS is an example of a TDD system.

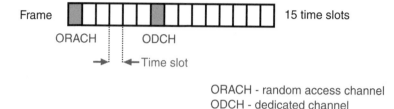

ORACH - random access channel
ODCH - dedicated channel

Figure 10.18 Time slot assignments in TDD option.

for random access and one for data. This relaying protocol can be used to improve the efficiency of the network by increasing the range of high-rate data services [3].

The advantages of the TDD network include flexibility for asymmetric links, compatibility with networks of limited radio resources, and availability of reciprocal channel measurements for better open-loop power control [3].[7] In contrast, the TDD network suffers these disadvantages: discontinuous transmission on the radio links, which creates pulse interference that affects local and neighbor cells, and slower power control because of fewer TPC symbols.

10.4 Multicarrier CDMA Option

The evolution path for CDMA IS95 includes an enhancement known as cdma2000(1x) and leads to the multicarrier option of the 3GPP standard. The cdma2000(1x) standard provides the same flexible attributes (physical channels, logical channels, and spreading codes) that are found in the 3GPP multicarrier option but keeps the narrow modulation bandwidth associated with CDMA IS95.[8] The 3GPP multicarrier system replicates the cdma2000(1x) data channel concept three times to produce the nominal 5-MHz bandwidth and leads to the common designation "cdma2000(3x)." Note that the aggregate spreading rate of the multicarrier scheme is three times the fundamental rate found in CDMA IS95 systems and is different from the 3GPP single-carrier option.

Both cdma2000(1x) and 3GPP multicarrier systems are designed to be compatible with existing CDMA IS95 networks. The systems share the same radio spectrum and, as such, require similar air interface performance, which is outlined next.

---

7. This concept uses the information in the previous slot's secondary sync channel to train the mobile phone's receiver [3]. The training algorithm compares the received data with the expected data and adjusts the receiver to maximize the cross-correlation result.

8. The narrowband spreading rate for both CDMA IS95 and cdma2000(1x) is 1.2288 Mcps.

10.4.1 Forward Link for the Multicarrier Option

The multicarrier forward-link modulator shown in Figure 10.19 resembles the forward-link modulator used in CDMA IS95 systems. It performs the following operations: FEC encoding, block interleaving, data scrambling, rate matching, Walsh covering, complex spreading, RF translation, and amplification. To minimize interaction and allow separate power control, the forward-link modulator processes each physical channel independently.

The forward link consists of several different logical channels, as shown in Table 10.6. Cdma2000(1x) and the 3GPP multicarrier option introduce several new channels to improve capacity, throughput, and flexibility. These standards also adopt the logical channels from CDMA IS95, including the common pilot channel, sync channel, paging channel, and traffic-fundamental channel.

The auxiliary pilot channels provide dedicated pilot signals for beamforming applications and improved spatial diversity, thereby assisting demodulation of high data rate signals. It is coded with a quasi-orthogonal Walsh function to avoid interference.

The broadcast and forward common control channels communicate to the mobiles within the cell coverage of the base station. The forward common control channel carries system overhead information and dedicated messages useful for the data link and network protocol layers.

The quick paging channel extends the standby time of the CDMA mobile telephone. It does that by communicating a simple, uncoded message about

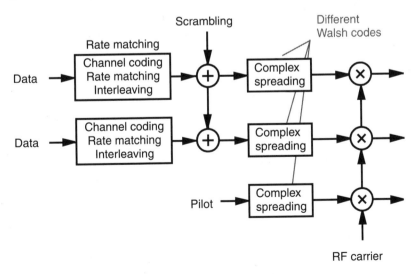

Figure 10.19 Block diagram of the multicarrier CDMA forward-link modulator.

Table 10.6
Forward-Link Channel Parameters in Multicarrier CDMA [32]

Channel	Data Rate (Kbps)	Channel Coding	Processing Gain
Pilot	—	None	—
Auxiliary pilot		None	Up to 512
Sync	1.2	Rate 1/2	1024
Paging	4.8, 9.6	Rate 1/2	256, 128
Broadcast	4.8, 9.6, 19.2	Rate 1/2	256, 128, 64
Quick paging	2.4, 4.8	None	512, 256
Common power control	19.2	None	64
Common assignment	9.6	Rate 1/2	128
Common control	9.6, 19.2, 38.4	Rate 1/4	128, 64, 32
Dedicated control	9.6	Rate 1/4	128
Fundamental			
Rate set 1	1.2, 2.4, 4.8, 9.6	Rate 1/2	1024, 512, 256, 128
Rate set 2	1.8, 3.6, 7.2, 14.4	Rate 1/2	682.7, 341.3, 170.7, 85.3
Supplemental	1.5, 2.7, 4.8N Turbo	Rate 1/4	819.2, 455.1, 256, 128/N

the common control and paging channels, which indicates whether to receive (and demodulate) the encoded channels. That reduces processing time and thereby improves standby time.

The common power control channel directs the power level transmitted by the mobile telephones within the cell coverage of the base station. It consists of multiple subchannels for multiple reverse-link channels and replaces power control bits that were punctured into the data stream in CDMA IS95.

The common assignment channel provides fast reverse-link channel assignments and thus supports random access packet data. The forward dedicated control channel informs the user of transmission and signaling information.

The fundamental data channels support the basic rate sets associated with existing vocoder standards, 8 Kbps (rate set 1) and 13 Kbps (rate set 2). Higher rates are possible when additional or supplemental data channels are added. Supplemental channels offer fixed rate (9.6 Kbps) or adjustable rate (1.2 to 307 Kbps for 1x and 1.0 Mbps for 3x) service. The fixed rate service supports up to seven additional supplemental channels, while the adjustable rate service—possible because of the variable spreading factors—supports just two supplemental channels.

The higher data rates provide less processing gain and thus less protection against interference. To combat that, turbo coding can be used above data rates of 14.4 Kbps. In addition, control data can be arranged in 5-ms frames instead of the standard 20 ms.

The forward-link logical channels in cdma2000(1x) and 3GPP multicarrier map directly to physical channels, which are separated by different extended orthogonal codes.

10.4.2 Reverse Link of the Multicarrier Option

The reverse-link modulator, shown in Figure 10.20, differs noticeably from the CDMA IS95 reverse-link modulator because it supports multiple physical channels and uses continuous transmission. It still performs the same basic operations, FEC encoding, rate matching, block interleaving, Walsh covering, complex spreading, RF translation, and amplification. But unlike the forward-link modulator, it combines the signals at baseband before RF translation and amplification to minimize hardware in the mobile telephone.

The reverse link consists of several logical channels, as listed in Table 10.7. The list includes the CDMA IS95 access and fundamental traffic channels and introduces several new channels for increased capacity and flexibility.

The reverse pilot channel provides a true reference for coherent detection at the base station and thus promises performance improvement of up to 3 dB. Its characteristics match those of the forward-link pilot signal. Also, the transmitter uses the reverse pilot channel power level as a reference power for the other physical channels.

The enhanced access channel performs tasks similar to those of the access channel: It initiates communications and responds to directed messages. To provide greater flexibility, the enhanced access channel offers three modes: basic access, power controlled, and reservation access.

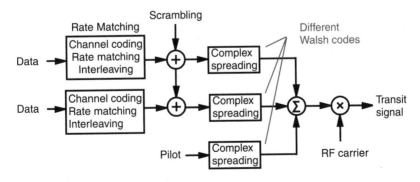

Figure 10.20 Block diagram of the multicarrier CDMA reverse-link modulator.

Table 10.7
Reverse-Link Channel Parameters for Proposed 3GPP Multicarrier Option [32]

Channel	Data Rate (Kbps)	Channel Coding	Processing Gain
Reverse pilot	—	None	—
Access	4.8	Rate 1/3	256
Enhanced access	9.6, 19.2, 38.4	Rate 1/4	32, 64, 128
Common control	9.6, 19.2, 38.4	Rate 1/4	32, 64, 128
Dedicated control	9.6, 14.4	Rate 1/4	128, 85.33
Fundamental			
Rate set 1	1.2, 2.4, 4.8, 9.6	Rate 1/3	128
Rate set 2	1.8, 3.6, 7.2, 14.4	Rate 1/2	85.33
Supplemental	1.5, 2.7, 4.8N	Rate 1/4	819.2, 455.1, 256/N
	1.8, 3.6, 7.2N	Rate 1/4	682.7, 341.3, 170.7/N

The reverse common control channel carries user and signaling information to the base station when the traffic channels are inactive. Otherwise, the dedicated control channel communicates that information.

The fundamental channel supports the basic rate sets 1 and 2. The supplemental channels (a maximum of two is allowed) provide higher data rates with variable spreading factors. With certain radio configurations, the data rate of each supplemental approaches 1.0 Mbps and an aggregate rate of 2.0 Mbps. In practice, a high data rate on the reverse-link is almost never needed because the radio communication is usually asymmetric, with much more data on the forward link.

The reverse-link data channels can also use turbo coding above 14.4 Kbps. Requests for higher data rates are made through the reverse access channels.

10.4.3 Power Control

A new outer loop power control algorithm, based on FER, is also introduced for cdma2000(1x) and the 3GPP multicarrier option [3]. The algorithm monitors the FERs for all the mobiles in the cell coverage area and adjusts the transmit power level for the different users to optimize performance. It minimizes the near-far effect and relies on performance measurements to improve capacity. To achieve the tighter control required by the new outer loop algorithm, finer power control steps are introduced. At the same time, more dramatic power control steps are also needed to support multi-rate design.

References

[1]   Dahlman, E., et al., "UMTS/IMT-2000 Based on Wideband CDMA," *IEEE Communications Magazine*, Sept. 1998, pp. 70–80.

[2]   Ojanpera, T., and R. Prasad, eds., *Wideband CDMA for Third Generation Mobile Communications*, Norwood, MA: Artech House, 1998.

[3]   Ojanpera, T., and R. Prasad, "An Overview of Air Interface Multiple Access for IMT-2000/UMTS," *IEEE Communications Magazine*, Sept. 1998, pp. 80–95.

[4]   Grieve, D., "RF Measurements for UMTS," *IEEE Colloq. on UMTS Terminals and Software Radio*, Apr. 1999, pp. 4/1–4/10.

[5]   Russel, T., *Signaling System #7*, New York: McGraw-Hill, 1998.

[6]   Dina, E. H., and B. Jabbari, "Spreading Codes for Direct Sequence CDMA and Wideband CDMA Cellular Networks," *IEEE Communications Magazine*, Sept. 1998, pp. 48–54.

[7]   Adachi, F., M. Sawahashi, and K. Okawa, "Tree-Structured Generation of Orthogonal Spreading Codes With Different Length for Forward Link of DS-CDMA Mobile Radio," *Electronics Letters*, Vol. 33, No. 1, Jan. 1997, pp. 27–28.

[8]   Gold, R., "Maximum Recursive Sequences With 3-Valued Recursive Cross-Correlation Functions," *IEEE Trans. on Information Theory*, Vol. IT-4, Jan. 1968, pp. 154–156.

[9]   Kasami, T., "Weight Distribution Formula for Some Class of Cyclic Codes," *Coordinated Science Lab*, Univ. of Illinois, Urbana, Technology Report R-285, Apr. 1966.

[10] Lin, S., and D. J Costello, Jr., *Error Control Coding: Fundamentals and Applications*, Englewood Cliffs, NJ: Prentice Hall, 1983.

[11] Berrou, C., A. Glavieux, and P. Thitimajshima, "Near Shannon Limit Error-Correcting Coding And Decoding: Turbo Codes," *Internat'l Conf. on Communications 1993*, Geneva, May 1993, pp. 1064–1070.

[12] Berrou, C., and A. Glavieux, "Near Optimum Error Correcting Coding and Decoding: Turbo Codes," *IEEE Trans. on Communications*, Vol. 44, No. 10, Oct. 1996, pp. 1261–1271.

[13] Wang, C. C., "On the Performance of Turbo Codes," *IEEE Proc. Military Communications Conf.*, 1998, Vol. 3, pp. 987–992.

[14] Robertson, P., "Illuminating the Structure of Code and Decoder of Parallel Concatenated Recursive Systematic (Turbo) Codes," *Proc. Globecom '94*, San Francisco, Dec. 1994, pp. 1298–1303.

[15] Bahl, L., et al., "Optimal Decoding of Linear Codes for Minimizing Symbol Error Rate," *IEEE Trans. on Information Theory*, Vol. 20, Mar. 1974, pp. 284–287.

[16] Hall, E. K., and S. G. Wilson, "Design and Analysis of Turbo Codes on Rayleigh Fading Channels," *IEEE J. on Selected Areas of Communications*, Vol. 16, No. 2, Feb. 1998, pp. 160–174.

[17] Taub, H., and D. L. Schilling, *Principles of Communication Systems,* Reading, MA: Addison-Wesley, 1995.

[18] Gallagher, M. D., and R. A Snyder, *Mobile Telecommunications Networking With IS-41*, New York: McGraw-Hill, 1997.

[19] Prasad, N. R., "GSM Evolution Towards Third Generation UMTS/IMT2000," *IEEE Conf. Personal Wireless Communications*, 1999, pp. 50–54.

[20]   Digital cellular telecommunications system (Phase 2+), High Speed Circuit Switched Data (HSCSD), service description; Stage 2, GSM 03.34.

[21]   ETSI, TS 03 64 v5.1.0 (197-11), Digital cellular telecommunications system (Phase 2+); General Packet Radio Service (GPRS); Overall description of the GPRS Radio interface: stage 2 (GSM 03.64 version 5.1.0).

[22]   Prasad, N. R., "An overview of General Packet Radio Services (GPRS)," *Proc. WPMC'98*, Yokosuka, Japan, Nov. 1998.

[23]   3GPP Technical Specification, "Physical Channels and Mapping of Transport Channels Onto Physical Channels (FDD)," TS 25.211, June 1999 (adjusted for new chip rate).

[24]   3GPP Technical Specification, "Multiplexing and Channel Coding (FDD)," TS 25.213, June 1999 (adjusted for new chip rate).

[25]   Kaplan, E., ed., *Understanding GPS: Principles and Applications*, Norwood, MA: Artech House, 1996.

[26]   Pickholtz, R. L., D. L. Schilling, and L. B. Milstein, "Theory of Spread-Spectrum Communications—A Tutorial," *IEEE Trans. on Communications*, Vol. 30, No. 5, May 1982, pp. 855–884.

[27]   Viterbi, A. J., *CDMA: Principles of Spread Spectrum Communications*, Reading, MA: Addison-Wesley, 1995.

[28]   Golomb, S. W., *Shift Register Sequences*, Aegean Park Press, 1992.

[29]   Fredricsson, S., "Pseudo-Randomness Properties of Binary Shift Register Sequences," *IEEE Trans. on Information Theory*, Vol. IT-B, Oct. 1967, pp. 619–621.

[30]   Proakis, John G., *Digital Communications*, New York: McGraw-Hill, 1995.

[31]   Hewlett Packard application note 1298, "Digital Modulation in Communication Systems—An Introduction."

[32]   TIA/EIA Proposal, "Physical Layer Standard for CDMA2000 Spread Spectrum Systems," IS-2000, Mar. 1999.

11

Advanced CDMA Mobile Radios

Many factors are spurring wireless communications, including phenomenal subscriber growth, the convergence of mobile radios and computers, and the potential services offered by next-generation CDMA systems. But there are also many obstacles that are forcing advances to the mobile radio. These advances target improved portability (smaller, lighter units with extended battery life), multimode operation (to better support roaming), and increased utility (additional software applications).

Advances in mobile radio technology are occurring in every major technology area. More powerful computers are enabling sophisticated algorithms and new applications. More integrated and flexible RF systems are reducing the size, weight, and current consumption of the RF transceiver, and more efficient PAs are extending battery life.

This chapter summarizes key advances in the areas of the digital system, the RF receiver, the RF transmitter, and the frequency synthesizer.

11.1 Advances in Digital Signal Processing

The role of the DSP in the mobile radio is growing. It is being asked to support next-generation applications, enable sophisticated digital receivers, implement reconfigurable architectures, and replace traditional analog functions. That is possible and actually highly desirable because of continuing improvements in DSP performance.

The massive use of digital signal processing algorithms is key to the implementation of advanced CDMA mobile radio applications. The requirements placed on the signal processing algorithms will increase dramatically in the years ahead as higher data rates and more sophisticated modulation, demodulation, and networking strategies become prevalent. Smart antennas, adaptive multipath equalization, synchronization, and variable bit-rate coding will all require significant improvements in DSP performance.

11.1.1 DSP Performance

The mobile phone places a tremendous burden on the DSP, a burden that is growing as the number of DSP applications escalates and the allocated execution times shrink. Fortunately, the DSP computer is developing at a phenomenal rate, as shown in Figure 11.1.

Advances in integrated circuit technology, architectures, and algorithms are producing DSPs with more computing power that dissipate less power, a critical point for battery-powered electronics. Advances in integrated circuit technology are relentless; amazingly, CMOS transistor density continues to double every 18 months [2]. That improves DSP functionality and, at the same time, keeps power dissipation at reasonable levels because, as transistors scale to smaller dimensions, the logic voltage levels and interconnect capacitance shrink. The power dissipation of a CMOS logic gate (P_d) is given by

$$P_d = C_{load} V_p^2 f_{clk} \qquad (11.1)$$

where C_{load} is the load capacitance presented to the logic gate, V_p is the supply voltage, and f_{clk} is the clock rate for the gate. Equation (11.1) clearly shows

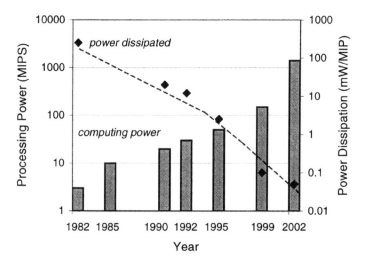

Figure 11.1 The computing power of the DSP is growing at an almost exponential rate while power dissipation is shrinking [1].

that power dissipation decreases with the square of the supply voltage. The combination of lower supply voltage and lower interconnect capacitance in advanced CMOS technologies neatly balances out the higher clock rates and greater number of gates integrated into the DSP. The surprising overall effect is shrinking power dissipation.

Next-generation DSP computers can support several new algorithms, including asynchronous pilot acquisition, advanced digital receivers, 1/2 rate vocoders, turbo decoders, and MPEG decoders [3]. The algorithms can be hardware or firmware based, depending on the data rate and the flexibility requirements.

Reconfigurable logic is another possibility; it combines the benefits of high-speed processing and flexibility. Unfortunately, this option requires several times more gates than hardwired solutions and therefore is limited to low-gate count functions.

11.1.2 Improvements to the Digital Receiver

One of the motivations for next-generation wireless communications is increased capacity, which is needed to serve a user population that is growing at an amazing rate (see Figure 1.1). But additional users bring new challenges to the mobile radio receiver. Next-generation digital receivers most likely will contain the functions outlined here to solve some of those challenges.

In direct-sequence spread-spectrum communications, the received signal at baseband is described in Section 2.1.1 as

$$r(t) = pn(t)Ad(t) + n'(t) + i'(t) \qquad (11.2)$$

where $pn(t)$ is the pseudorandom modulating waveform, A is the amplitude of the message waveform, $d(t)$ is the message signal with bipolar values ±1, $n'(t)$ is thermal noise, and $i'(t)$ is interference. It is received by a conventional digital receiver, based on the correlator shown in Figure 11.2(a), which produces an output signal equal to

$$pn(t)r(t) = pn^2(t)Ad(t) + pn(t)n'(t) + pn(t)i'(t) \qquad (11.3)$$

Ideally, the PN sequences at the transmitter and the receiver are synchronized so that $pn^2(t) = 1$ and, after integration, the bit energy is collapsed back to its original bandwidth R. Additionally, any received interference $i(t)$ is spread by the correlator to the relatively wide bandwidth W, and its effect is lowered by the processing gain of the system, W/R.

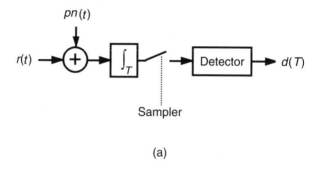

(a)

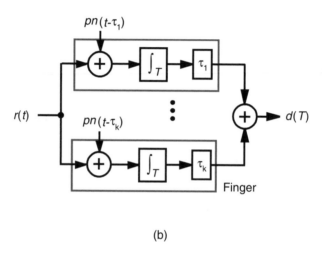

(b)

Figure 11.2 Digital receiver structures: (a) correlator and (b) Rake receiver.

The received signal, unfortunately, is a composite of many phase-shifted and distorted versions of the transmitted waveform, known as multipath rays. Multipath rays smear the transmitted data pulses and produce ISI. In narrowband TDMA single-user systems such as GSM, multipath delay spread is mitigated by an equalizer [4]. In wideband spread-spectrum systems, the Rake receiver (described in Section 5.2.4) shown in Figure 11.2(b) constructively combines the multipath rays.

The Rake receiver, however, is optimal only for a single-user system with binary data. In practice, other users share the CDMA channel and degrade system performance. This was first pointed out in the simple SNR analysis in Section 2.1.2. There, the other users were treated as noise and the SNR was expressed as

$$SNR \approx \frac{S}{\sum\limits_{n=0}^{k-1} S_n} \approx \frac{pn^2(t)Ad(t)}{pn(t)i'(t)} \qquad (11.4)$$

where thermal noise is neglected and S_n is the received power intended for other users. At the mobile radio, the received power for each user may or may not be equal. In fact, the received power is equal when the signals are transmitted from a single base station. That situation is less likely, though, when the received signals are from multiple base stations, because the transmitted signals propagate via different radio channels. Consequently, the near-far problem [5] can arise. The interference due to other user signals is generally labeled as multiple access interference.

Another potential problem is in-band interference. In-band interfering signals are unaffected by RF channel select filters. They are, however, suppressed by the correlator in the digital receiver, although that is limited to the processing gain of the spread-spectrum system [6]. Note that the interfering signals can be troublesome for next-generation systems, where higher data rates use lower spreading factors (and provide less processing gain).

The problems of in-band interference, multiple access interference, and the near-far effect are addressed by two advanced digital receiver concepts, interference rejection and multiuser detection.

11.1.2.1 Interference Rejection

A spectral filtering technique is used to remove in-band interfering signals. The technique relies on notch filters, which greatly reduce the effect of the interfering signal but also introduce distortion [7]. These digital filters typically are programmed using estimation methods and are limited to a few percent of the spread-spectrum bandwidth.

The estimation method is based on tapped delay line structures [7]. The tapped delay line operates on half chip-rate ($T_c/2$) samples of the desired received waveform. The data are uncorrelated for the CDMA signal because of the noiselike spreading signal $pn(t)$ but are correlated for the narrowband interfering signal. As such, linear prediction [8] can be used to estimate the next sample,[1] as shown in the simple single-sided transversal filter in Figure 11.3(a).

It is also possible to use a two-sided transversal filter like the one shown in Figure 11.3(b), although that increases processing delay. Here, past and future samples are used to estimate the current sample.

---

1. Linear prediction is described in Section 4.2.2 for speech signals.

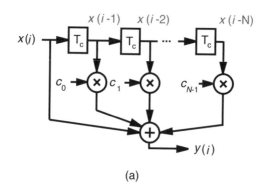

(a)

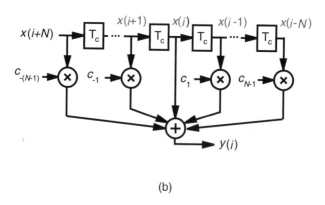

(b)

Figure 11.3 Interference rejection techniques: (a) single-sided tapped delay line, (b) transversal filter, (c) lattice filter, and (d) decision feedback equalizer (DFE).

The optimum tapped weights in the filter are computed using a least mean square (LMS) algorithm that minimizes the MSE between the received signal, $r(t)$, and the expected signal, $pn(t)Ad(t)$. The LMS algorithm is an approximation to the Wiener-Hopf equation [9]:

$$\mathbf{W}_{opt} = \mathbf{R}^{-1}[r(t)]\mathbf{R}[r(t),\, pn(t)Ad(t)] \qquad (11.5)$$

where $\mathbf{R}^{-1}[r(t)]$ is the inverse autocorrelation matrix for the received signal and $\mathbf{R}[r(t),\, pn(t)Ad(t)]$ is the cross-correlation matrix of the received signal to the expected signal. In practice, that approximation can be implemented by a number of approaches, including the Widrow-Hoff [10], Levinson-Durbin [4], and Burg [11] algorithms.

The relatively slow convergence of the LMS algorithms has given way to several other transversal filter structures, the most popular being the lattice

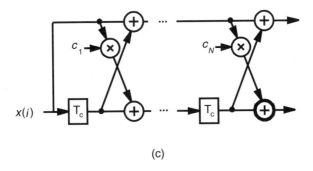

(c)

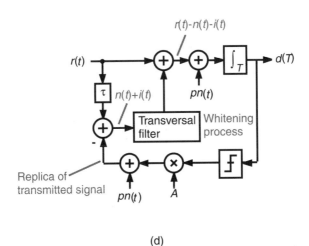

(d)

Figure 11.3 (continued).

structure [7], shown in Figure 11.3(c). In that structure, each section of the lattice filter converges independently.

Another alternative to the transversal filter is the decision feedback equalizer [4, 9], shown in Figure 11.3(d). It strives to "whiten" only the noise and the interference, not the desired signal. In principle, the DFE subtracts the desired signal from the received signal before processing by a "whitening" filter. The DFE relies on the output of the receiver to generate a replica of the desired waveform. The drawback of this approach is that it relies on the receiver output, which, if incorrect, can propagate errors [7].

11.1.2.2 Multiuser Detection and the Near-Far Problem

The effects of multiple access interference and imperfect power control can be reduced using multiuser detection techniques. These techniques detect other user signals and mitigate their effects, thereby improving the SNR of the desired signal [12–15].

The optimal communications receiver consists of a bank of correlators, assigned to each possible transmitted signal, and a joint detector. In CDMA communication systems, the number of correlators in the optimal receiver is at least equal to the number of users. That is unrealistic because the complexity of multiuser detector (MUD) schemes grows exponentially with the number of users [16].

In the above receiver, it was assumed that all the received signals are orthogonal, a relationship that breaks down in practice because signals are rarely received synchronously [17]. That leads to measurable cross-correlation between the multiple access signals and the desired signal, with

$$R(n) = \lim_{T \to \infty} \frac{1}{T} \int_{-T/2}^{T/2} pn(t) pn_n(t + \tau) dt \qquad (11.6)$$

where T is the correlation length, $pn(t)$ are the multiple access spreading codes, and τ is the time misalignment. An increase in the cross-correlation result leads to greater probability of error in the detection process [12, 17]. Additionally, multiuser detection techniques require signals with an adequate SNR for accurate channel estimation. That is problematic in CDMA systems that continually adjust the transmitted power to compensate near-far effects.

Another important consideration in the correlator receiver is its length, T. Ideally, that should cover the spreading code and any excess delay due to multipath propagation. However, that can be extensive in certain environments.

Because the optimal receiver is extremely complex, a suboptimal, less complicated approach is needed [12]. As a result, these approaches are designed to operate on the strongest signals because those signals contribute the greatest multiple access interference (MAI) and provide reliable estimates. Also, removing the strongest MAI signals combats the near-far problem.

In the multistage and the decision feedback detectors, the MAI from the strongest signals is detected and subtracted from the received signal. The detection process occurs at the bit rate after despreading. In the successive interference canceler, estimates of the strongest MAI signals are made and subtracted from the received signal. By contrast, this method uses sub-chip rate samples.

The successive interference canceler is the simplest approach and yields the best results [18]. Its structure consists of two or three canceling circuits before the Rake receiver, as shown in Figure 11.4. The circuits remove the effects of the strongest signals, which usually are selected a priori. The common

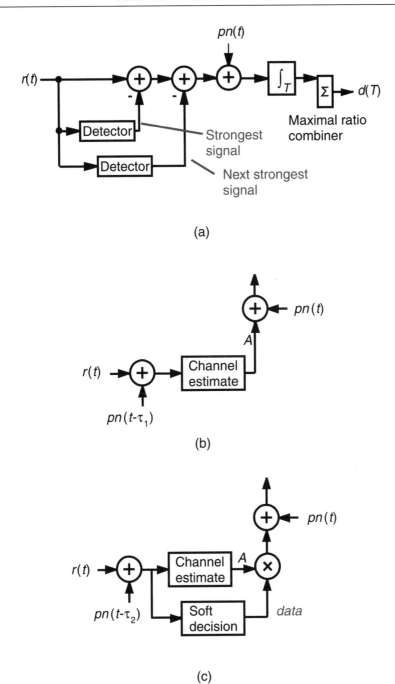

(a)

(b)

(c)

Figure 11.4 Digital receiver with successive interference canceler: (a) block diagram, (b) applied to pilot signal, and (c) applied to data.

pilot, for example, is a prime candidate because it is the strongest transmitted signal. Note that this signal is already estimated by the Rake receiver.[2] Additional candidate signals are the high data rate signals transmitted in next-generation systems. These signals will be transmitted at higher power levels to maintain the wireless link and E_b/N_o ratio. This information is available from the radio resource layer of the protocol stack.

11.2 Advanced RF Receivers

Higher integration, lower power dissipation, and lower cost provide strong motivation for research into advanced RF receiver architectures. The classic problem in RF receiver design is the image response of the downconverter mixer. Armstrong's invention of the super heterodyne receiver in the 1930s elegantly solved that problem. It consisted of a tunable front-end filter (to filter out the image signal), tunable first LO, combined with a relatively high, fixed, first IF frequency to ease the bandwidth requirements of that filter. The resulting receiver, which is used extensively in CDMA mobile radios, has outstanding selectivity and sensitivity.

A low-loss, tunable front-end filter is difficult to realize in solid-state form and is bulky in its classic mechanical configuration. It is possible to replace the tunable filter with a fixed filter with a very sharp cutoff and relatively wide bandwidth, which is implemented in a straightforward manner with a SAW device. The SAW filter has its own problems, including physical size, high insertion loss, and cost. Hence, there is a need for improved RF receiver architectures that eliminate the need for image rejection filtering.

11.2.1 Image Rejection Techniques

Most of the alternative architectures solve the image signal problem geometrically, through the use of orthogonal signal techniques. There are essentially two different image rejection techniques, one introduced by Hartley [19], shown in Figure 11.5(a), and the other introduced by Weaver [20], shown in Figure 11.5(b). Both start with the same basic structure, which consists of two mixers driven by orthogonal LO signals, $\cos \omega_{LO} t$ and $\sin \omega_{LO} t$. If the input signal consists of two signals, the desired $A_d \cos \omega_D t$ and its image $A_i \cos \omega_I t$, the output of the first mixer after low-pass filtering—which removes the sum products produced by the mixer—is

---

2. In the Rake receiver, the channel is estimated from the pilot signal (Section 5.2.2).

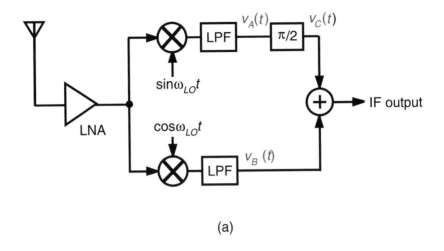

(a)

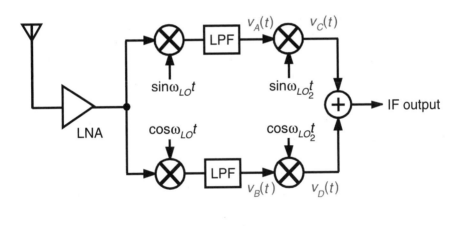

(b)

Figure 11.5 Image reject receivers: (a) Hartley and (b) Weaver.

$$v_A(t) = (A_d \cos \omega_d t + A_i \cos \omega_i t) \sin \omega_{LO} t \qquad (11.7)$$

$$= \frac{A_d}{2}[\sin(\omega_d - \omega_{LO})t] + \frac{A_d}{2}[\sin(\omega_i - \omega_{LO})t]$$

Similarly, the output of the second mixer is

$$v_B(t) = (A_d \cos\omega_d t + A_i \cos\omega_i t) \cos\omega_{LO} t \qquad (11.8)$$

$$= \frac{A_d}{2}[\cos(\omega_d - \omega_{LO})t] + \frac{A_i}{2}[\cos(\omega_i - \omega_{LO})t]$$

At this point, notice that the image and the desired signals are both downconverted to the same frequency but are orthogonal to each other. The Hartley architecture takes the output from the first mixer (driven by $\sin\omega_{LO}t$) and shifts it by 90 degrees ($\pi/2$). That differentiates the downconverted signal from the downconverted image signal and produces the signal[3]

$$v_C(t) = -\frac{A_d}{2}\cos(\omega_d - \omega_{LO})t + \frac{A_i}{2}\cos(\omega_i - \omega_{LO})t \qquad (11.9)$$

When that result is subtracted from the output of the second mixer, $v_B(t)$, the image signal is canceled and the desired signal is obtained.

The Weaver architecture takes the outputs from each mixer and downconverts them a second time to a final IF frequency, using another pair of mixers. In some cases (e.g., TV tuners), the signal is first downconverted to dc and then upconverted to a fixed IF frequency [21]. In the more typical case, however, the output of the second set of mixers (once again after suitable filtering) is at baseband and is given by

$$v_C(t) = -\frac{A_d}{2}\cos(\omega_d - \omega_{LO} + \omega_{LO2})t + \frac{A_i}{2}\cos(\omega_i - \omega_{LO} - \omega_{LO2})t \quad (11.10a)$$

$$v_D(t) = \frac{A_d}{2}\cos(\omega_d - \omega_{LO} + \omega_{LO2})t + \frac{A_i}{2}\cos(\omega_i - \omega_{LO} - \omega_{LO2})t \quad (11.10b)$$

The result is the same as the Hartley image reject structure, when the second output is subtracted from the first, that is, the image signal is nulled.

The Weaver architecture is the basis of the so-called wideband IF downconversion architecture [22], where the first LO is a fixed downconversion, and the second downconversion is tuned to the desired IF frequency. A schematic of this approach is shown in Figure 11.6.

Both the Hartley and the Weaver downconverters have the desirable feature of eliminating the image response as well as the image noise, an important practical advantage. However, although straightforward in principle, each has some significant disadvantages in practice. First of all, it is important to note that the utility of the scheme depends on the degree of image rejection that can be achieved by the architecture.

---

3. This assumes high-side injection mixer, where $\omega_{LO} > \omega_{RF}$, and results in $\omega_d - \omega_{LO} < 0$.

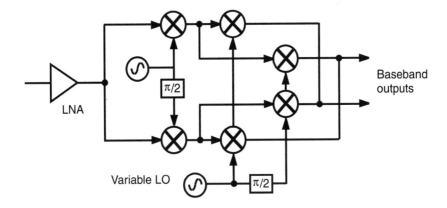

Figure 11.6 Wideband IF downconversion architecture [22].

Imperfect image rejection arises from gain and phase mismatches between the two paths of the downconverter. If the relative power gain mismatch is given by $\Delta A/A$, and the phase error in radians is denoted by $\Delta\theta$, then the image rejection ratio (the ratio of the image gain to the desired signal gain) is given by [23]

$$IRR \approx \frac{\left(\frac{\Delta A}{A}\right)^2 + \Delta\theta^2}{4} \qquad (11.11)$$

which is an approximation to (8.3) and (9.38). Typical results for an integrated circuit process are better than 30 dB, indicating gain differences less than approximately 1% and phase differences less than 3 degrees.

In most cases, this architecture is employed to allow a relatively low IF frequency, potentially eliminating one extra stage of downconversion. However, an image rejection ratio of 30 to 50 dB is inadequate for most mobile wireless applications, due to the high level of in-band interferers. Image rejection downconverters are more practical in dual-band receivers. One of the advantages of the image reject receiver is that it can select either the desired signal or the image signal, depending on the sign of the summing block. That useful feature can be employed to receive two widely separated bands (say at 900 MHz and 1,900 MHz) with a single downconverter and a relatively high IF frequency (500 MHz in this case). The approach is shown in Figure 11.7 [24]. Although the image rejection of the receiver itself is rather poor—still only 30 to 40 dB—the overall image rejection is improved by the front-end duplex filter, which provides an additional 40–50 dB attenuation.

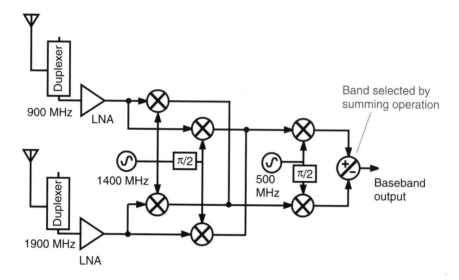

Figure 11.7 Dual-band receiver using switchable image rejection to select the desired band of downconversion [24].

11.2.2 Direct Conversion Receivers

An alternative to the image rejection approach is the use of homodyne, or direct downconversion, techniques, as shown in Figure 11.8. The desired signal is downconverted to baseband in a single step, eliminating the problem of image responses completely. The architecture is highly amenable to completely monolithic implementations of the entire receiver and is the focus of many research efforts. The desired channel is simply extracted with an appropriate low-pass filter at the output prior to A/D conversion. This architecture is

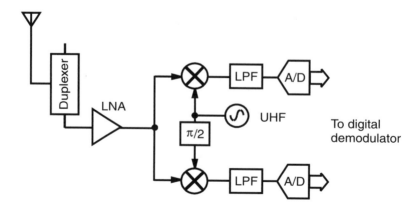

Figure 11.8 Homodyne architecture.

elegantly simple in concept and solves most of the problems associated with classical heterodyne approaches, but it introduces a myriad of problems of its own.

The first problem is associated with the choice of the IF frequency—dc. Any dc offsets in the system will be indistinguishable from the desired signal. Figure 11.9 summarizes a variety of sources of pernicious dc offsets.

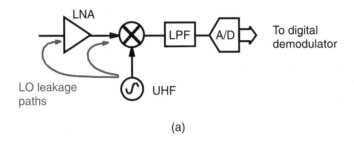

(a)

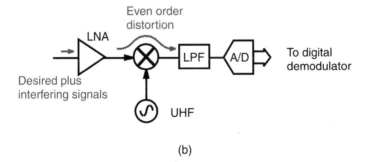

(b)

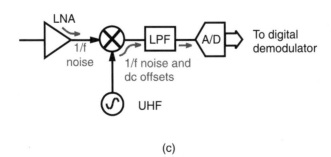

(c)

Figure 11.9 Sources of dc errors in homodyne receivers: (a) LO leakage, (b) even-order harmonic distortion, and (c) 1/f noise leakage.

First, note that the desired signal and the LO are centered at the same frequency. Any leakage of the LO to the input of the mixer or LNA will downconvert right on top of the desired signal, as illustrated in Figure 11.9(a).

Second, note that any even harmonic distortion in the mixer can leak through to the IF port. The problem here is that the entire input signal—the desired signal, all the other signals in the channel, and the interferers—will experience even-order distortion and downconvert to dc, as illustrated in Figure 11.9(b). That is further complicated because any interference due to second-order distortion has a bandwidth twice that of the original signal.

Finally, all the circuits associated with the downconversion process exhibit a nonnegligible systematic dc offset, as well as $1/f$ noise, as illustrated in Figure 11.9(c). These problems are inconsequential in most heterodyne receivers, where the final IF frequency is still well above dc, but in the homodyne architecture, they can be crippling.

In theory, such limitations can be overcome through a variety of well-known dc suppression techniques. One possibility is ac coupling of the output, in which the signal is high-pass filtered with suitably large capacitors to eliminate the offsets, noise, and interference at dc. Unfortunately, that removes a portion of the desired signal but does not eliminate all the noise and interference. At most, only a small percentage of the bandwidth can be notched without affecting the performance of spread-spectrum communications.

Differentiating the signal before digitization and then re-forming the signal by integrating digitally can also mitigate the problem of dc offset in the homodyne receiver. An implementation of this technique, shown in Figure 11.10, uses an adaptive delta-modulator. The feedback loop forces the output

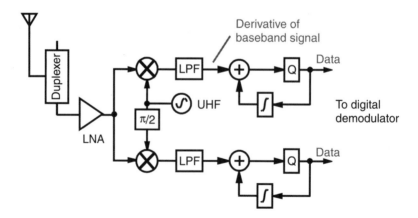

Figure 11.10 Differentiation of the output signal of a homodyne receiver to reduce dc offset effects [25].

of the integrator to be equal to that of the input signal and therefore requires that the input to the integrator be the derivative of the input signal [25].

In TDD systems, such as GSM, the dc offset and other system nonlinear effects can be measured during a training sequence. That is not possible, however, in full-duplex systems like CDMA IS95 and next-generation CDMA systems.

One of the advantages of the homodyne receiver compared to the hetero-dyne Hartley and Weaver architectures is its relative insensitivity to mismatch effects in the two branches of the downconverter. That is because the image signal is simply the desired signal itself. Furthermore, any gain mismatch can be corrected digitally.

Another problem associated with the homodyne architecture is LO radia-tion. Leakage from the LO port of the mixer to the RF port can couple to the antenna and radiate, possibly corrupting the received signal of nearby users. This is not implausible, because the mixer is driven by a strong LO signal and LO-RF isolation is limited. In addition, the radiated LO signal can be re-received and cause problems.

The LO radiation problem can be partially overcome by the use of a harmonic mixer. In Figure 11.11(a) [26], the mixer "mixes" on both the positive and negative going waveforms of the LO, achieving an effective doubling of the frequency. Alternatively, a two-level mixing scheme can be employed, as shown in Figure 11.11(b) [27]. The LO frequency is now precisely one-half the desired frequency, which is easily filtered by the duplex filter. Additionally, a fully differential structure will exhibit extremely low second harmonic distor-tion of the LO, minimizing the output of the mixer at the desired frequency. To work effectively, the harmonic mixers must be driven in half-quadrature with respect to each other or at 45 degrees phase shift.

Another drawback of the homodyne architecture is the dynamic range requirements imposed on the baseband filter and gain stages. The baseband filters must provide extra stopband rejection, or, alternatively, the A/D convert-ers must cover a wider dynamic range, since the IF SAW filter has been removed. In practice, the dynamic range burden on the early filter stages means those stages are often realized with purely passive elements [28].

11.2.3 Digital IF Receivers

In the classical super heterodyne architecture, the digital IF receiver shifts much of the analog signal processing to the digital system. It does that by sampling the received signal at the IF frequency and performing I/Q demodulation digitally, as shown in Figure 11.12. The digital receiver exploits advances in

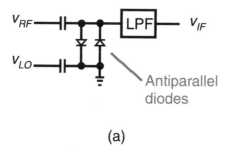

(a)

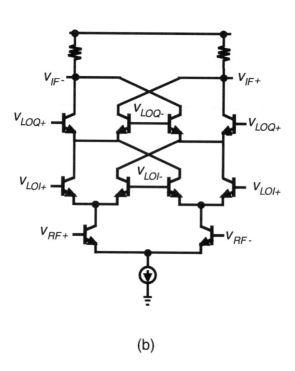

(b)

Figure 11.11 Harmonic mixer for direct downconversion applications: (a) antiparallel diode version and (b) two-level mixing approach.

A/D converter performance and digital technology to lower power dissipation and achieve higher integration.

One of the major impediments to this approach is the performance of the A/D converter. The problem of digitization becomes increasingly difficult as the IF frequency rises or dynamic range requirements grow. Progress in the field of A/D converters has been incremental at best over the last 30 years, as shown in Figure 11.13, with improvements on the order of 1 bit of resolution every six to eight years [29].

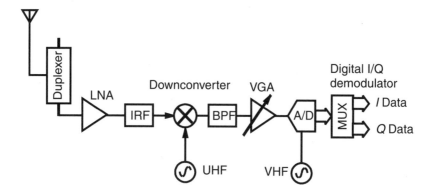

Figure 11.12 IF sampling in the digital receiver relies on a high-performance A/D converter.

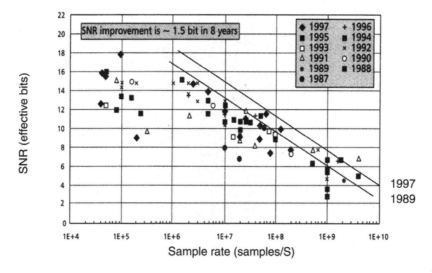

Figure 11.13 Progress of A/D converter performance (*From:* R. H. Walden, "Performance Trends for Analog-to-Digital Converters," *IEEE Communications Magazine,* © 1999 IEEE).

It is not necessary to sample the IF signal at twice the IF frequency. That is because the information is contained in the modulation or envelope of the signal, not the carrier. As such, it is acceptable to sample the IF signal at a far lower rate, based on the bandwidth of the modulation. Furthermore, it is convenient to sample the IF signal at four times the bandwidth of the modulation. That produces an output pattern of I, Q, −I, −Q, which greatly simplifies digital demodulation, because a simple demultiplexer yields I and Q information with perfect balance.

Subsampling at a frequency less than the carrier frequency dramatically lowers the requirements on the A/D converter. Unfortunately, subsampling receivers suffer from noise aliasing. Any noise present at the input to the A/D converter is folded to the bandwidth of $f_s/2$. This is typically limited by an antialiasing filter. But the bandwidth of the antialiasing filter must be wide enough to pass the IF signal—including the carrier—to prevent distortion prior to sampling. As a result, subsampling by a factor of m multiplies the downconverted noise power by a factor $2m$. Furthermore, the error due to sampling jitter (see Section 6.1.2) depends on the IF carrier frequency, not the modulation frequency.

11.2.4 Comparison of Advanced RF Receiver Architectures

Current CDMA IS95 mobile radios primarily employ the super heterodyne architecture. Table 11.1 compares the super heterodyne architecture with the advanced receiver architectures, direct conversion, and digital IF.

11.3 Advanced RF Transmitters

The conflicting goals of linearity and power-added efficiency in transmitter PAs set a fundamental limit on the performance of RF transmitters. Next-generation CDMA communication systems place a greater burden on the RF transmitter, particularly the PA, because the proposed modulation schemes produce a carrier envelope with a larger peak-to-average ratio. In addition, to achieve maximum capacity in spread-spectrum communication systems, it is important to keep the received power at the base station roughly constant. In

Table 11.1
Comparison of Prominent RF Receiver Architectures

Architecture	Benefits	Challenges
Heterodyne	Proven architecture, high selectivity, wide dynamic range	Integration, frequency plan
Digital IF	Excellent I/Q demodulation, low power, fewer analog circuits	Power control, dynamic range, frequency plan, sampling process
Direct conversion	Simple architecture, integration, adaptive	Self-mixing, second-order distortion, low frequency noise

a typical wireless environment, the mean output power is less than the peak level and is always changing with time. As a result, the average efficiency of the PA often is poor.

Achieving high efficiency and linearity across a broad range of output power levels is the goal of many advanced RF transmitter architectures and improved PA topologies. Such architectures and topologies are the subject of active research and are described next.

11.3.1   Direct Conversion Transmitters

The direct conversion transmitter is similar to the direct conversion receiver. It is a highly integrated solution that directly converts baseband orthogonal signals to RF frequency, as shown in Figure 11.14.

There are several drawbacks to this architecture, many of them similar to those associated with the direct conversion receiver. The output of the PA is a digitally modulated signal centered at the RF carrier frequency that is the same as the LO frequency. If even a small fraction of the PA output is injected into the LO, the LO will acquire the modulation of the transmit signal, and the modulation accuracy will be hopelessly compromised.

This well-known phenomenon of oscillator design is known as injection locking—the LO becomes injection locked to the output of the PA [30]. The magnitude of the effect depends on how close the injection locking signal's frequency is to that of the (formerly) free-running oscillator. In the case of a direct upconversion transmitter, the two frequencies are identical, and the problem can be severe.

There are some possible solutions to this dilemma. One solution is to use two LOs, each far removed in frequency from the desired signal, and then

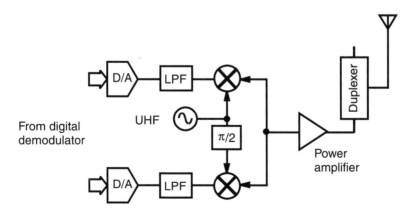

Figure 11.14 Direct upconversion architecture.

multiply the two signals together to obtain the sum or difference frequency for the required carrier. This approach is shown in Figure 11.15. Each LO is far enough removed in frequency that the output frequency has no chance to "pull" either of the oscillators.

A second problem with direct conversion upconverters is related to the same gain and phase mismatch problems as the downconversion architecture. Orthogonal errors create some signal "leakage" from the I to the Q path and vice versa. The magnitude of the signal leakage is approximately [23]

$$\frac{P_{leakage}}{P_{desired}} \approx \frac{\left(\frac{\Delta A}{A}\right)^2 + \Delta \theta^2}{4} \tag{11.12}$$

where $\Delta A / A$ is the relative power gain imbalance between the two channels and $\Delta \theta$ is the phase imbalance between the two channels. Typically, a suppression of greater than -40 dB is required. That is easier to achieve at lower frequencies than at microwave frequencies, which is another reason for the unpopularity of direct conversion techniques in the transmitter.

11.3.2 SSB Techniques

A standard mixer generates both sum and difference products, one of which is wanted while the other is removed to prevent spurious problems. The unwanted product is typically removed by filtering or by using SSB mixing techniques.

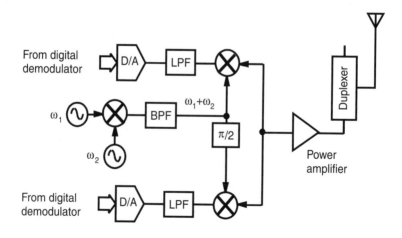

Figure 11.15 Offset mixer architecture for realization of direct upconversion transmitter.

An SSB mixer, shown in Figure 11.16, is based on the same principles as the image reject mixer. For an input signal $A \cos \omega t$, the output of the phase shifter is $A \sin \omega t$ and the output of the first mixer is

$$v_A(t) = A \sin \omega_{IF} t \sin \omega_{LO} t \tag{11.13}$$

$$= \frac{A}{2} \cos(\omega_{IF} - \omega_{LO})t - \frac{A}{2} \cos(\omega_{IF} + \omega_{LO})t$$

Similarly, the output of the second mixer is

$$v_B(t) = A \cos \omega_{IF} t \cos \omega_{LO} t \tag{11.14}$$

$$= \frac{A}{2} \cos(\omega_{IF} - \omega_{LO})t + \frac{A}{2} \cos(\omega_{IF} + \omega_{LO})t$$

When the outputs of the two mixers, $v_A(t)$ and $v_B(t)$ are combined, the result is the difference product, referred to as the lower sideband. When the output of the second mixer, $v_B(t)$, is subtracted from the output of the first mixer, $v_A(t)$, the sum product, known as the upper sideband, is formed. Sideband suppression is analyzed using (11.11).

Another SSB mixing technique uses the frequency translation loop [31], shown in Figure 11.17. It consists of a phase detector, two low-pass filters, a VCO, and an offset mixer. The system functions as a PLL, with the mixer used to frequency shift the RF signal to IF. When the loop is in synchronization mode, the output of the mixer is phase-locked to the IF input signal and is at one of two frequencies, either $f_{LO} - f_{RF}$ or $f_{RF} - f_{LO}$. The polarity of the phase detector output selects the frequency of the VCO and, in turn, the output frequency of the mixer.

The frequency translation loop greatly reduces spurs in the output of the transmitter, since the RF signal is formed by the VCO, not an upconversion mixer. Furthermore, this architecture is suitable for dual band transmitters.

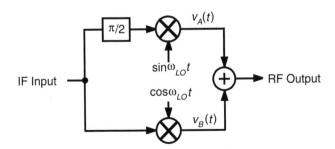

Figure 11.16 Upconversion SSB mixer.

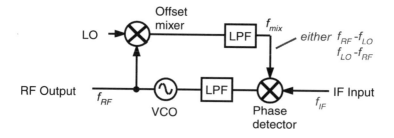

Upper sideband mixer
$$f_{RF} = f_{IF} + f_{LO}$$
when $f_{IF} > f_{mix}$, f_{RF} *increases*

lower sideband mixer
$$f_{RF} = f_{LO} - f_{IF}$$
when $f_{IF} > f_{mix}$, f_{RF} *decreases*

Figure 11.17 Frequency translation loop [31].

11.3.3 Predistortion Techniques for Amplifier Linearization

One of the simplest conceptual approaches for the improvement of linearity in the transmitter PA is the technique of predistortion. A typical PA exhibits gain compression at high input powers, which results in AM-AM conversion, and often exhibits excess phase shift at high input powers, which results in AM-PM conversion. Together, those effects create distortion and intermodulation in the high-power output of the amplifier.

If the input to the PA could be predistorted with the inverse of its own nonlinearity, the overall effect of the nonlinearity could be canceled out. This is shown conceptually in Figure 11.18. The predistortion circuit would ideally compensate for both the gain and the phase nonlinearity of the amplifier circuit and would therefore exhibit both gain and phase expansion at the high input power levels.

Although straightforward in principle, the predistortion approach suffers from several practical drawbacks. First of all, it is impossible to track precisely the effects of temperature, process, and power supply variations on PA nonlinearity. The problem is difficult because the levels of acceptable distortion are very low, and a small drift between the PA and the predistortion circuit can create substantial out-of-band interference.

It is also true that the predistortion could be performed at baseband using digital techniques if the appropriate transformation function for the predistorter were known in advance. That technique, illustrated in Figure 11.19, is known as adaptive predistortion [32]. The distortion through the amplifier is measured

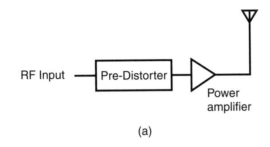

(a)

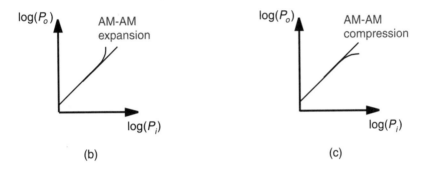

(b)   (c)

Figure 11.18 Predistortion applied to PA linearization: (a) schematic diagram, (b) predistortion transfer function, and (c) PA response.

periodically, and its AM-AM and AM-PM conversion is calculated. The data is then fed to the DSP, which provides the I and Q signals for the baseband upconverter, and the DSP predistorts the output of the modulator to provide the necessary linearizing response. Several different versions of adaptive predistortion have been developed.

The obvious practical problems with the predistortion concept naturally lead to an exploration of more robust techniques for achieving the desired goal. The traditional approach to linearization of a nonlinear analog system is feedback. With appropriate feedback, the transfer function of the predistortion circuit naturally tracks the highly variable transfer function of the nonlinear PA.

An example of a possible feedback approach for a PA is illustrated in Figure 11.20(a). A linear operational amplifier supplies the necessary predistortion of the signal in a precise manner, in response to the difference between the (distorted) output signal and the desired input signal. This straightforward approach has the obvious drawback that an operational amplifier with the required bandwidth and output drive capability simply does not exist at micro-

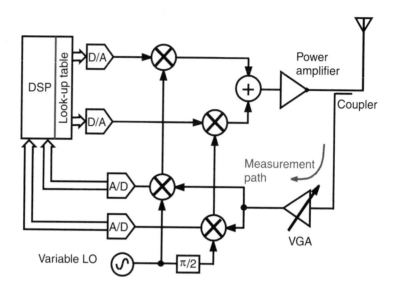

Figure 11.19 Adaptive predistortion employs a measurement of the output waveform to produce the necessary input compensation.

wave frequencies. Furthermore, the phase shift associated with a typical PA is highly variable, making stability difficult to achieve under a wide range of conditions.

Providing the feedback at lower frequencies by downconverting the amplified signal is one possibility, as shown in Figure 11.20(b). The first drawback is that the downconversion mixers have to be as linear as the desired output signal. That is not a problem in most cases, because only a small portion of the output signal is required for feedback purposes, easing the linearity requirements of the mixer considerably.

A larger problem is that of excess phase shift through the combination of PA, mixer, and low-pass filter. In general, the phase shift is hard to control at microwave frequencies and varies, depending on the power level. An additional variable phase shift is, therefore, necessarily added to the mixer to ensure stability under all conditions. That phase shift must be carefully controlled over process, temperature, and power supply variations. The feedback approach is also prone to problems associated with amplifier saturation and rapid changes in output VSWR [33].

Digital modulation techniques typically require upconversion of both the I and Q baseband signals. As a result, feedback typically is applied to both paths of the PA inputs, with a technique known as Cartesian feedback, shown in Figure 11.20(c). Cartesian feedback has been an active research topic over many years [34], but it has not achieved widespread adoption because of

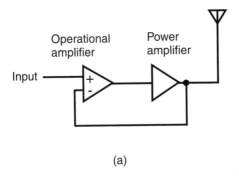

(a)

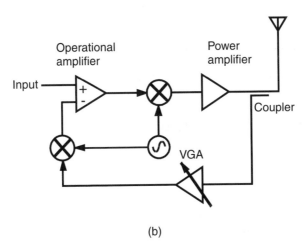

(b)

Figure 11.20 Amplifier linearization using feedback: (a) simplified view of feedback linearization approach, (b) use of frequency-translating downconverter to achieve linearization, and (c) Cartesian feedback applied to provide both gain and phase correction.

the inherent difficulties in applying feedback across a large and complicated microwave circuit. It is also possible to digitize the feedback signal and perform the feedback using the DSP. That has the advantage of being able to alter the phase shift adaptively in order to maintain stability. However, the approach suffers from the same drawback that plagues all feedback control systems, that is, the bandwidth of the system is limited by the loop delay. Hence, an all-digital approach to Cartesian feedback will have to await the arrival of dramatically faster DSPs and A/D converters.

11.3.4 Feedforward PAs

The myriad of problems associated with the predistortion approaches—both open-loop and feedback—point to an opportunity for alternative solutions.

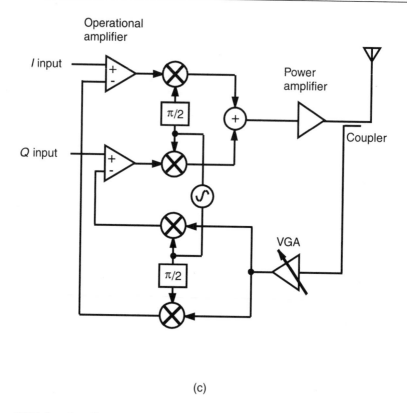

(c)

Figure 11.20 (continued).

Rather than predistorting the input signal, it might be more effective to measure the nonlinearity of the PA, subtract the error generated by the nonlinearity from the ideal signal, amplify the difference, and then subtract the difference from the amplifier output. That approach, although seemingly complicated, has been used successfully for many years to linearize satellite traveling wave tube amplifiers (TWTAs) and is known as the feedforward approach [35]. It is illustrated schematically in Figure 11.21.

Feedforward techniques for amplifier linearization actually predate the use of feedback techniques. Both were developed by Black in the 1930s to solve the problem of linearization for telephone network repeater amplifiers [36]. A close examination of Figure 11.21 reveals the reason that feedback techniques quickly supplanted feedforward techniques for most lower frequency applications. First of all, the gain and phase matching between the two input paths of the subtractor circuit must be precisely matched to achieve acceptable cancelation of the distortion products. Second, the gain of the error amplifier must precisely track the gain of the PA itself. Finally, the phase shift through

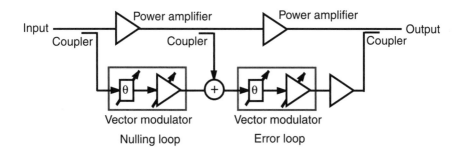

Figure 11.21 Feedforward predistortion of nonlinear PAs.

the final phase shift network and hybrid coupler must precisely track the gain and phase shift of the PA.

Despite those apparent obstacles, the use of feedforward approaches has several adherents, although it is typically employed in base station and higher frequency circuits, where power efficiency is less important than absolute linearity.

11.3.5   Linearized PAs With Nonlinear Circuits

The techniques described in the preceding sections rely on linearization of a nearly linear amplifier to achieve the desired specifications. The two techniques described in this section, envelope elimination and restoration (EER) and linear amplification with nonlinear components (LINC), achieve linear amplification through fundamentally nonlinear processes. Their advantage is potentially much higher efficiency without a sacrifice in linearity.

EER (also known as the Kahn technique [37]) relies on the principle that the PA operates in its most power-efficient mode at its peak output power, for example, 78.5% in class B mode. However, the peak is rarely achieved under normal operation if the power supply is fixed. That suggests the strategy of varying the power supply of the amplifier in response to variations in the input waveform. In the limit, the amplifier operates in a pure switching mode (highly nonlinear and efficient), and all the variation in the output envelope is provided by the variation in the power supply voltage. This approach is shown in Figure 11.22. In theory, the overall efficiency of the technique is limited by the efficiency of the dc-dc converter supplying the power supply to the PA and the efficiency of the PA itself.

Several potential drawbacks with this approach need to be considered. First, the phase shift between the two branches of the amplifier must be carefully matched; any difference in delay will cause distortion in the resulting signal. The phase shift associated with limiting stages has a high degree of amplitude

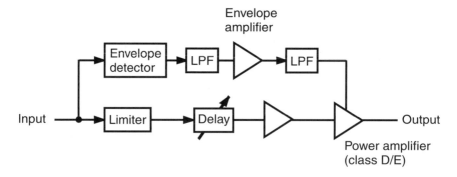

Figure 11.22 EER technique.

dependence, so AM-PM conversion in the amplifier stage needs to be replicated in the power supply stage.

Second, the power supply needs to accommodate variations in the envelope, which can occur at roughly the chip rate in a direct-sequence spread-spectrum system. Efficient switching power supplies that operate at those frequencies have yet to be developed, although there do not appear to be any fundamental technological obstacles to their development. In addition, the response of the envelope detector and the power supply together now set the overall linearity of the circuit, and careful attention must be paid to the linear design of those circuits.

A possible alternative to a pure EER system is to operate the amplifier in the class A/class AB mode and simply rely on variations in the power supply to improve the efficiency, rather than rely on the power supply itself to supply the needed envelope variations. This approach does not result in as dramatic an improvement in power-added efficiency as the EER technique, but it minimizes the need for precise phase alignment in the two branches of the amplifier. This approach, which is shown in Figure 11.23, can lead to dramatic improvements in power-added efficiency over the full range of output power in a typical CDMA environment [38]. Because the load line of the amplifier does not change, it is advantageous to change both the drain voltage and the bias current in response to variations in the input amplitude to achieve the best possible efficiency [39]. This optimized strategy is illustrated in Figure 11.24.

The concept of outphasing amplification has a long history (dating to Chireix in the 1930s [40]). The technique has been revived under the rubric of LINC and applied to a variety of wireless applications. The concept itself is simple: Two amplifiers are operated with constant envelope input signals (hence, very power efficient), and their outputs are summed to produce the desired signal. The desired envelope and phase variation at the output is

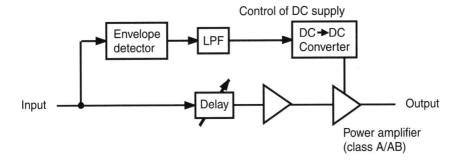

Figure 11.23 Variable power supply for tracking the envelope variations of the input signal. The amplifier remains in the class A mode over its entire range of operation.

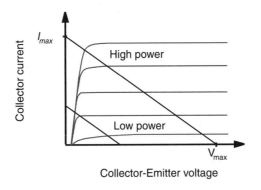

Figure 11.24 Optimized load-line strategy for best performance of class A PA with tracking power supply.

obtained by varying the relative phases between the two signals, as shown schematically in Figure 11.25.

The desired phase variation between the two amplifiers originally was obtained using analog techniques; now digital approaches are more typically used.

Despite its apparent attractiveness, the LINC approach has several disadvantages that have limited its applicability. The first is that the power typically is summed with a hybrid power-combining network, as shown in Figure 11.25(a). That portion of the power delivered to the hybrid that is not delivered to the antenna is dissipated in the 50Ω terminating resistor. As a result, the amplifier achieves its peak operating efficiency only at maximum output power, and its efficiency decreases linearly as the output power decreases. Such efficiency behavior is comparable to that of a class A amplifier, which is known to have

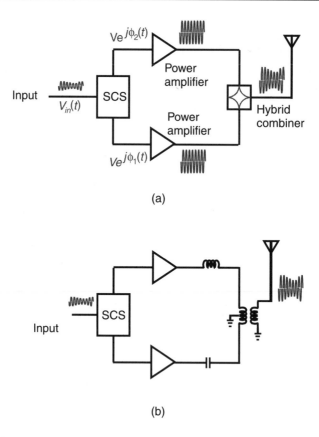

(a)

(b)

Figure 11.25 Illustration of outphasing amplifier concept: (a) use of hybrid power combiner for signal summation and (b) Chireix power-combining technique.

very poor overall efficiency. Of course, the peak efficiency of the LINC approach is much higher than that of the class A amplifier, but it would be desirable to do even better.

One problem with the previous power-combining approach is the power wasted in the power-combining network. It is not desirable to simply connect the output of the two amplifiers together, because the output phase of one amplifier will affect the output phase and impedance of the other amplifier. As a result, the load impedance presented to each amplifier appears highly reactive over a large portion of the cycle, harming the efficiency. A partial solution can be achieved using the so-called Chireix power-combining technique, illustrated in Figure 11.25(b) [41]. Two impedance transformers are added to improve the efficiency. The added susceptance (which is inductive in one branch and capacitive in the other branch) cancels out the varying susceptance seen by each amplifier at one particular output power.

A second problem associated with the LINC approach is the gain and phase mismatch associated with the two branches of the amplifiers. Any mismatch between the two can lead to severe intermodulation and distortion [42]. Typical requirements for CDMA applications are on the order of less than 0.3-degree phase mismatch and less than 0.5-dB gain mismatch, a near impossibility in most practical cases. As a result, several compensation or calibration schemes have been proposed [43]. Those techniques have not achieved wide application, because of their inherent complexity and lack of flexibility.

11.4 Advanced Frequency Synthesizers

The classical PLL architecture suffers from a variety of limitations, which make its use for mobile wireless applications less than ideal. The most significant of those limitations is the tradeoff between frequency spacing, which must be equal to the reference frequency and is therefore a small fraction of the output frequency, and the loop bandwidth, which should be as large as possible to minimize phase noise. Because the loop bandwidth is limited to roughly no more than a few times the reference frequency, it is difficult to produce both narrow frequency spacing and broad loop bandwidth.

Fractional-N PLL architectures are one approach to overcome that limitation. The frequency division inside the loop can take on noninteger values. At least in principle, that allows the spacing between the output frequencies to be less than the reference frequency, allowing for a wider loop bandwidth and reduced phase noise. A variety of differing approaches to the implementation of this circuit are available; Figure 11.26(a) shows one example [43].

The implementation of a fractional-N synthesizer is straightforward and relies on varying the modulus of a frequency divider between two adjacent integers, for example, 10 and 11. Such frequency dividers are called dual-modulus frequency dividers. If a 10/11 divider is operated in divide-by-10 mode half the time and divide-by-11 mode the other half, then the average division ratio will be 10.5. More generally, if the divider divides the output of the VCO by N for J cycles and divides the output of the VCO by $(N + 1)$ for K cycles, looping between the two modes constantly, there will be $[NJ + (N + 1)K]$ VCO pulses for every $(J + K)$ reference pulses. Then,

$$(NJ + (N + 1)K)\tau_{VCO} = (J + K)\tau_{REF} \qquad (11.15)$$

or

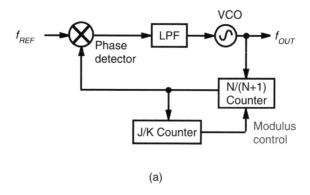

(a)

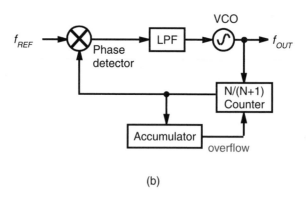

(b)

Figure 11.26 Fractional-N PLL architectures: (a) J/K counter and (b) accumulator.

$$f_{VCO} = \left(N + \frac{K}{J+K}\right)f_{REF} \qquad (11.16)$$

which is a noninteger fraction of the input or reference frequency.

Another approach employs a K-bit accumulator at the output of the dual-modulus divider, as shown in Figure 11.26(b). The accumulator is preset to a value F, where the division ratio is $[N + F/(2^k - 1)]$. Each time the accumulator overflows, the dual-modulus divider divides by $(N + 1)$ instead of N.

This approach to "fractional-N synthesis" is remarkably simple and elegant. However, it suffers the major drawback of introducing spurious frequency modulation of the VCO output. Those spurs can be identified by noting that the divider does not really divide by a fractional value—it divides by either a smaller- or a larger-than-desired value. During the time the divider is dividing by a smaller-than-desired integer, say, 10 in the earlier example, a phase error between the output and the input begins to accumulate. The phase error reaches a maximum at the point where the master counter has counted to J pulses. Then the counter begins to count by a larger-than-desired value, and the phase error begins to decrease. The average phase error (over time) is zero, but the time-varying output of the phase detector modulates the VCO input, as mediated by the loop filter. The period of this waveform is

$$\tau_{mod} = \tau_{VCO}(J + K) \tag{11.17}$$

which will create spectral sidebands around the desired frequency at integral multiples of the resulting frequency, $f_{VCO}(J + K)$.

The spectral sidebands can be significant in a wireless receiver, because of the problem of reciprocal mixing (outlined in Section 7.4.2). There are, however, several different approaches to eliminate the problem.

The first approach is to note that the accumulating phase error is deterministic in the sense that it is precisely known for a given fractional division ratio. Subtraction of the phase error by a compensating analog circuit is employed in many commercial fractional-N synthesizers in an attempt to eliminate the problem. This approach is shown in Figure 11.27 [44]. The drawback of this approach is that the matching requirements of the analog compensation circuitry and the PLL are difficult to achieve in practice. Additionally, the noise generated by the compensation circuitry must be extremely low, because it directly modulates the VCO control line.

A second approach is to randomize the phase error in some manner, so that the periodic modulation of the VCO control line is replaced with a randomly varying control signal, as shown in Figure 11.28 [45]. If the energy of the phase error is not increased and the average division ratio is unchanged, then the total spectral energy that was originally in the discrete sidebands is smeared out over a much wider bandwidth.

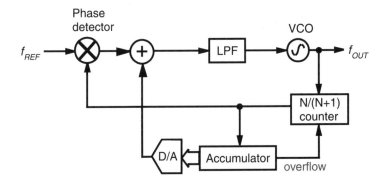

Figure 11.27 Analog compensation of errors in fractional-N PLL.

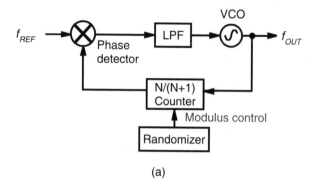

(a)

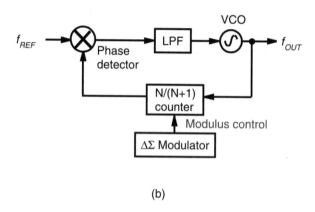

(b)

Figure 11.28 Elimination of discrete spurious tones in a fractional-N synthesizer output through (a) randomization of the modulus and (b) $\Delta\Sigma$ modulation of the modulus.

The randomization process is usually accomplished by a pseudorandom noise generator. The randomization can be chosen so that the average value of the modulus is correct, but the division ratio is varied randomly between N and $N + 1$. Most pseudonoise sequences have flat spectral properties, generating broad-bandwidth spectral sidebands. Alternatively, the pseudorandom binary data can be high-pass noise-shaped so that the spurious spectral sidebands are outside the band of interest of the synthesizer [46]. A circuit that generates this particular sequence of digital data is known as a $\Delta\Sigma$ modulator (see Section 6.2.4). This approach is particularly useful, because it places the noise energy of the resulting synthesized output well away from the desired output frequency.

References

[1] Frantz, G. A., "The Future of Digital Signal Processing," presentation by Texas Instruments, Mar. 1999.

[2] Poor, H. V., and G. W. Wornell, eds., *Wireless Communications*, Upper Saddle River, NJ: Prentice Hall, 1998.

[3] Sikora, T. "MPEG Digital Video-Coding Standards," *IEEE Signal Processing Magazine*, Vol. 14.5, Sept. 1997, pp. 82–100.

[4] Proakis, J, G., *Digital Communications*, New York: McGraw-Hill, 1995.

[5] Cooper, G. R., and C. D. McGillen, *Modern Communications and Spread Spectrum*, New York: McGraw-Hill, 1986.

[6] Iltis, R. A., and L. B. Milstein, "Performance Analysis of Narrow-Band Interference Rejection in DS Spread-Spectrum Systems," *IEEE Trans. on Communications*, Vol. COM-32, No. 11, Nov. 1984, pp. 1169–1177.

[7] Milstein, L. B., "Interference Rejection Techniques in Spread Spectrum Communications," *IEEE Proc.*, Vol. 76, No. 6, June, 1988, pp. 657–671.

[8] Ketchum, J. W., and J. G. Proakis, "Adaptive Algorithms for Estimating and Suppressing Narrow-Band Interference in PN Spread-Spectrum Systems," *IEEE Trans. on Communications*, Vol. COM-30, No. 5, May 1982, pp. 913–924.

[9] Frerking, M. E., *Digital Signal Processing in Communication Systems*, Boston: Kluwer Academic Publishers, 1994.

[10] Widrow, B., and M. Hoff, "Adaptive Switching Circuits," *IRE WESCON Convention Record*, Pt. 4, 1960, pp. 96–104.

[11] Burg, J. P., "Maximum Entropy Spectral Analysis," reprinted in D. B. Childers (ed.), *Modern Spectral Analysis*, New York: IEEE Press, 1978, pp. 34–41.

[12] Due-Hallen, A., J. Holtzman, and Z. Zvonar, "Multiuser Detection for CDMA Systems," *IEEE Personal Communications*, Apr. 1995, pp. 46–58.

[13] Schneider, K. S., "Optimum Detection of Code Division Signals," *IEEE Trans. on Aerospace and Electronic Systems*, Vol. AES-15, No. 1, Jan. 1979, pp. 181–185.

[14] Kohno, R., M. Hatori, and H. Imai, "Cancellation Techniques of Co-Channel interference in Asynchronous Spread Spectrum Multiple Access Systems," *Electronics and Communications*, Vol. 66-A, No. 5, 1983, pp. 20–29.

[15] Varanasi, M., and B. Aazhang, "Multistage Detection in Asynchronous Code-Division Multiple Access Communications," *IEEE Trans. on Communications*, Vol. 38, Apr. 1990, pp. 505–519.

[16] Lupas, R. and S. Verdu, "Linear Multiuser Detectors for Synchronous Code-Division Multiple-Access Channel," *IEEE Trans. on Information Theory*, Vol. IT-35, No. 1, Jan. 1989, pp. 123–136.

[17] Verdu, S., "Minimum Probability of Error for Asynchronous Gaussian Multiple-Access Channels," *IEEE Trans. on Information Theory*, Vol. IT-32, No. 1, Jan. 1986, pp. 85–96.

[18] Patel, P., and J. Holtzman, "Analysis of a Simple Successive Interference Cancellation Scheme in a DS/CDMA System," *IEEE J. of Selected Areas of Communications*, Vol. 12, No. 5, June, 1994, pp. 796–807.

[19] Hartley, R., "Modulation System," US Patent 1,666,206, Apr. 1928.

[20] Weaver, D. K., "A Third Method of Generation and Detection of Single-Sideband Signals," *IRE Proc.*, Vol. 44, Dec. 1956, pp. 1703–1705.

[21] Aschwanden, F., "Direct Conversion: How to Make It Work in TV Tuners," *IEEE Trans. on Consumer Electronics*, Vol. 42, No. 3, pp. 729–738.

[22] Rudell, J. C., et. al, "A 1.9 GHz Wideband IF Double Conversion CMOS Integrated Receiver for Cordless Telephone Applications," *ISSCC Dig. Tech. Papers*, Feb. 1997, pp. 304–305.

[23] Razavi, B., *RF Microelectronics*, Prentice Hall, 1998.

[24] Wu, S., and B. Razavi, "A 900 MHz/1.8 GHz CMOS Receiver for Dual-Band Applications," *IEEE J. of Solid-State Circuits*, Vol. 33, No. 12, Dec. 1998, pp. 2178–2185.

[25] Lindquist, B., and M. Isberg, "A New Approach to Eliminate the dc Offset in a TDMA Direct Conversion Receiver," 1993 43rd IEEE Vehicular Technology Conference, pp. 754–757.

[26] Itoh, K., et al., "A 40 GHz Band Even Harmonic Mixer With an Antiparallel Diode Pair," *IEEE MTT-S International Microwave Symp. Dig.*, 1991, pp. 879–882.

[27] J. Choma, "A Three-Level Broad-Banded Monolithic Analog Multiplier," *IEEE J. of Solid-State Circuits*, Vol. SC-16, No. 4, Aug. 1981, pp. 392–399.

[28] Chang, P., A. Rofougaran, and A. Abidi, "A CMOS Channel-Select Filter for a Direct Conversion Wireless Receiver," *IEEE J. of Solid-State Circuits*, Vol. 32, No. 5, May 1997.

[29] Walden, R. H., "Performance Trends for Analog-to-Digital-Converters," *IEEE Communications Magazine*, Feb. 1999, Vol. 37, No. 2, pp. 96–101.

[30] Kurokawa, K., "Injection Locking of Microwave Solid-State Oscillators," IEEE Proc., Vol. 61, Oct. 1973, pp. 1386–1410.

[31] Tham, J. L., et al, "A 2.7V 900MHz/1.9GHz Dual-Band Transceiver IC for Digital Wireless Communications," *IEEE Custom Integrated Circuits Conf.*, 1998, pp. 559–562.

[32] Mansell, A., and A. Bateman, "Adaptive Digital Predistortion Linearization," *Proc. Microwaves and RF Conf.*, London, Oct. 1996, pp. 270–275.

[33] Cripps, S., *RF Power Amplifiers for Wireless Communications*, Norwood, MA: Artech House, 1999.

[34] Petrovic, V., and C. Smith, "Reduction of Intermodulation Distortion by Means of Modulation Feedback," *IEE Conf. on Radio Spectrum Conservation Techniques*, Sept. 1983, pp. 44–49.

[35] Seidel, H., "A Microwave Feedforward Experiment," *Bell System Tech. J.*, Vol. 50, Nov. 1971, pp. 2879–2916.

[36] Black, H. S., "Translating System," US Patent 1,686,792, issued Oct. 29, 1928.

[37] Kahn, L. R., "Single-Sideband Transmission by Envelope Elimination and Restoration," *IRE Proc.*, Vol. 40, July 1952, pp. 803–806.

[38] Asbeck, P., et al., "Efficiency and Linearity Improvement in Power Amplifiers for Wireless Communications," *GaAs IC Symp.*, 1998, pp. 15–18.

[39] Saleh, A., and D. Cox, "Improving the Power-Added Efficiency of FET Amplifiers Operating With Variable Envelope Signals," *IEEE Trans. on Microwave Theory and Techniques*, Vol. 31, No. 1, Jan. 1983, pp. 51–56.

[40]   Chireix, H., "High Power Outphasing Modulation," *IRE Proc.*, Vol. 23, No. 11, Nov. 1935, pp. 1370–1392.

[41]   Raab, F. H., "Efficiency of Outphasing RF Power-Amplifier Systems," *IEEE Trans. on Communications*, Vol. COM-33, No. 10, Oct. 1985, pp. 1094–1099.

[42]   Casadevall, F., and J. Olmos, "On the Behavior of the LINC Transmitter," *IEEE Proc. Vehicular Technology Conf.*, 1990, pp. 29–34.

[43]   Razavi, B., *Monolithic Phase-Locked Loops and Clock Recovery Circuits*, New York: IEEE Press, 1996.

[44]   Phillips Semiconductor, Phillips RF/Wireless Communications Data Handbook, Phillips SA8025 Frequency Synthesizer, 1998.

[45]   Reinhardt, V., et al., "A Short Survey of Frequency Synthesizer Techniques," *Proc. 40th Annual Frequency Control Symp.*, May 1986, pp. 355–365.

[46]   Riley, T., M. Copeland, and T. Kwasniewsky, "Sigma-Delta Modulation in Fractional-N Synthesis," *IEEE J. of Solid-State Circuits*, Vol. 28, May 1993, pp. 553–559.

Glossary

ΔΣ	delta-sigma
1G	first generation
2G	second generation
3G	third generation
3GPP	3rd Generation Partnership Project
ACP	adjacent channel power
ACPR	adjacent channel power ratio
ACS	add-compare-select
A/D	analog to digital
ADPCM	adaptive differential pulse coded modulation
AFC	automatic frequency control
AGC	automatic gain control
ALC	automatic level control
ALU	arithmetic logic unit
AM	amplitude modulation
AMPS	analog mobile phone system
APCM	adaptive pulse code modulation
ARIB	Association of Radio Industry and Business
ARQ	automatic repeat request
ASIC	application-specific integrated circuit
AWGN	additive white Gaussian noise
BCJR	Bahl, Cocke, Jelinek, and Raviv
BER	bit error rate

BJT	bipolar junction transistor
BPSK	binary phase shift keying
CAI	communication air interface
CDMA	code division multiple access
CMOS	complimentary metal oxide semiconductor
CRC	cyclic redundancy check
D/A	digital to analog
DAM	diagnostic acceptability measure
dB	decibel
dBm	decibel milliwatt
dBW	decibel watts
dc	direct current
DFT	discrete Fourier transform
DLL	delay-locked loop
DNL	differential nonlinearity
DPCM	differential pulse coded modulation
DQPSK	differential QPSK
DRT	diagnostic rhyme test
DS	direct sequence
DSB	double sideband
DSP	digital signal processor
E&M	electricity and magnetism
EEPROM	electrical erasable/programmable read-only memory
EER	envelope elimination and restoration
ESN	electronic serial number
ETSI	European Telecommunications Standards Institute
EVM	error vector magnitude
EVRC	enhanced variable rate coder
F	noise factor
FDD	frequency division duplex
FDMA	frequency division multiple access
FEC	forward error correction
FER	frame error rate
FFT	fast Fourier transform
FH	frequency-hopped
FIR	finite impulse response
FM	frequency modulation

FPGA	field programmable gate array
Gbps	gigabit per second
GIPS	giga-instructions per second
GMSK	Gaussian minimum shift keying
GPRS	general packet radio protocol system
GPS	Global Positioning System
GSM	Global System for Mobile Communication
HBT	heterojunction bipolar transistor
HLR	home location register
HSCSD	high-speed circuit switched data
I	in-phase
IF	intermediate frequency
IIR	infinite impulse response
IMD	intermodulation distortion
IMD3	3rd-order intermodulation distortion
IMT	International Mobile Telecommunications
INL	integral nonlinearity
IP3	3rd-order intercept point
IRF	image reflect filter
IRR	image rejection ratio
IS	interim standard
ISDN	Integrated Services Digital Network
ISI	intersymbol interference
ISUP	ISDN user part
Kbps	kilobit per second
LAR	log-area ratio
LC	inductor-capacitor
LINC	linear amplification with nonlinear components
LNA	low-noise amplifier
LO	local oscillator
LP	linear prediction
LPC	linear prediction coder
LSP	line spectrum pair
LTP	long term prediction
MAC	medium access control
MAP	maximum a posteriori
MCU	microcontroller unit

MDS	minimum detectable signal
MESFET	metal semiconductor field effect transistor
MFLOPS	mega floating point operations per second
MIPS	mega-instruction per second
MLSE	maximum likelihood sequence estimation
MLSR	maximum length shift register
MOSFET	metal oxide semiconductor field effect transistor
MSE	mean-square error
MSK	minimum shift keying
MTP	message transfer part
NADC	North America Digital Cellular
NF	noise figure
NMOS	N-type metal-oxide semiconductor
NRZ	nonreturn-to-zero
NSC	nonsystematic convolutional
ODMA	opportunity driven multiple access
OMC	operation and maintenance center
OQPSK	offset quadrature phase shift key
OSI	Open Systems Interconnections
OVSF	orthogonal variable spreading factors
PA	power amplifier
PAE	power-added efficiency
PCM	pulse coded modulation
pdf	probability density function
PHEMT	pseudomorphic high electron mobility transistor
PHS	Personal Handyphone System
PIN	positive-intrinsic-negative
PLL	phase-locked loop
PM	phase modulation
PMOS	P-type metal-oxide semiconductor
PN	pseudorandom number
PSC	primary sync. code
psd	power spectral density
PTSN	Public Telephone Switching Network
Q	quadrature-phase
QAM	quadrature amplitude modulation
QCELP	Qualcomm code excited linear prediction

QoS	quality of service
QPSK	quaternary phase shift keying
RAM	random access memory
RF	radio frequency
RMS	root mean square
ROM	read only memory
RPE	regular pulse excitation
RSC	recursive systematic convolusional
S/H	sample/hold
SAW	surface acoustic wave
SEGSNR	segmented signal-to-noise ratio
SF	shape factor
SID	system identification number
SMS	short messaging Services
SNR	signal-to-noise ratio
SS7	Signaling System #7
SSB	single sideband
SSC	secondary sync. code
TDD	time division duplex
TDMA	time division multiple access
TFCI	transport format channel indicator
TIA	Telecommunications Industry Association
TPC	transport power control
TUP	telephone user part
TWTA	traveling wave tube amplifier
UHF	ultra high frequency
VCO	voltage controlled oscillator
VGA	variable gain amplifier
VHF	very high frequency
VLIW	very long instruction word
VLR	visitor location register
VLSI	very large scale integration
VSELP	vector summed excitation linear prediction
VSWR	voltage standing wave ratio

About the Authors

John B. Groe received a B.S. degree in electrical engineering in 1984 from California State University at Long Beach and an M.S. degree in electrical engineering in 1990 from the University of Southern California.

In 1980, he joined TRW, where he initially worked as an RF technician. From 1983 to 1991, he designed RF, phase-locked loop, and data converter integrated circuits for communications and radar applications. From 1991 to 1993, Mr. Groe was with Brooktree Corporation, where he designed integrated circuits for the automatic test equipment market. From 1993 to 1996, he was at Pacific Communication Sciences, Inc., where he designed integrated circuits for Japan's PHS communication system. In 1996, he joined Nokia, Inc., where he currently manages radio frequency integrated circuit design activities and directs research into advanced CDMA mobile radio architectures. Mr. Groe's research interests lie in the area of digital signal processing techniques to mitigate channel impairments and RF receiver nonlinear effects.

Mr. Groe is a senior member of IEEE. He has five U.S. patents and has several others pending, all in the area of wireless communications.

Lawrence E. Larson received a B.S. degree in electrical engineering in 1979 and a M. Eng. in 1980, both from Cornell University, Ithaca, New York. He received a Ph.D. in electrical engineering from the University of California—Los Angeles in 1986.

In 1980, Dr. Larson joined Hughes Research Laboratories, where he directed work on high-frequency InP, GaAs, and silicon integrated circuit development for a variety of radar and communications applications. From 1994 to 1996, he was at Hughes Network Systems, where he directed the development of RF integrated circuits for wireless communications applications.

He joined the faculty at the University of California—San Diego in 1996 and is the inaugural holder of the Communications Industry Chair. Dr. Larson has published over 120 papers and has received 21 U.S. patents. He is editor of the book, *RF and Microwave Circuit Design for Wireless Communications,* published by Artech House.

Dr. Larson was a co-recipient of the 1996 Lawrence A. Hyland Patent Award of Hughes Electronics for his work on low-noise millimeter wave HEMTs and the IBM General Managers Excellence Award. He is a fellow of the IEEE.

Index

Wideband CDMA for Third Generation Mobile Communications,
Tero Ojanperä and Ramjee Prasad, editors

Wireless Communications in Developing Countries: Cellular and Satellite Systems, Rachael E. Schwartz

Wireless Technician's Handbook, Andrew Miceli

For further information on these and other Artech House titles, including previously considered out-of-print books now available through our In-Print-Forever® (IPF®) program, contact:

Artech House
685 Canton Street
Norwood, MA 02062
Phone: 781-769-9750
Fax: 781-769-6334
e-mail: artech@artechhouse.com

Artech House
46 Gillingham Street
London SW1V 1AH UK
Phone: +44 (0)20 7596-8750
Fax: +44 (0)20 7630-0166
e-mail: artech-uk@artechhouse.com

Find us on the World Wide Web at:
www.artechhouse.com